Working Notes:

A Guide To Seahorse Diseases

Working Notes:

A Guide To Seahorse Diseases

A book from the Working Notes series

Sections authored by Martin Belli M.D.,
Clare Driscoll, Marc Lamont and others

Edited by Diane Bennett

Authored by   Martin Belli M.D.
Clare Driscoll
Marc Lamont
Pete Giwojna
Keith Gentry

August 2006

ISBN 1-904830-01-3

Published 2006 Creative Licence Publishers
Printed by CafePress.com in the United States of America

This book is dedicated to the group of friends collectively known as Seahorse.org. This group have put aside financial gains, petty disagreements and other distractions and have concentrated on the thing that binds us – the seahorse.

This book would not have been possible without the ideas, innovations and observations of the members of seahorse.org.

Acknowledgements

There are many people who have made important contributions to this guide. The authors and editor of this book have merely accumulated the knowledge from the home seahorse keeper, tested and refined this information, and presented it in a congruent format.

Additional investigative work and pathology have helped the understanding of this new and complicated subject, and hopefully this book will have the desired impact by providing a view, *'from our working notes'* of the knowledge we have accumulated on seahorse diseases.

This work would not be possible without the information produced by the many hobbyist researchers on whose work we have based our own research. It is the hobbyist who has helped make the advances in the knowledge of the seahorse, its requirements in a captive setting, and the identification and treatment of the diseases that can afflict seahorses.

We hope that it will be the hobbyist who will benefit most from the observations in this book.

Table of Contents

Foreword

This first volume of the Working Notes series is focused on the detection and treatment of disease and parasites in seahorses. The intended audience includes experienced seahorse hobbyists, fish stores, seahorse breeding and aquaculture facilities, aquariums, and zoos. As will be the case with subsequent volumes, this book is the result of the cooperative effort of committed seahorse keepers, pooling their accumulated knowledge and experience. It is a compilation and summary of the most recent information and recommendations in seahorse pathology, based on the research and experience of its seahorse hobbyist authors and the thousands of fellow hobbyists on Seahorse.Org.

This volume is not intended to be the "end-all" definitive work on seahorse disease. As the series title, Working Notes, suggests, it comprises our practical knowledge to date. However, the information contained in this volume consists of the most up to date information and recommendations currently available. The collection and communication of research and information via an extensive, international, Internet community of committed hobbyists is truly a new paradigm. It is one we feel will be more effective, accurate, useful and rapid than traditional methods of funded scientific research and publication, particularly where a specialized interest is being served.

To learn more about basic care of seahorses in the aquarium, to become part of this dynamic community, and to contribute to our ongoing base of knowledge of seahorse care, join Seahorse.Org.

Lisa Darmo, Ph.D.
July 18, 2005

Introduction

There are currently no relevant books on seahorse diseases.

The Working Notes series of books is an attempt to produce a book that is fresh and contains information relevant at the time of publishing. To this end the normal 'due process' involved in editing and proofing a book of this type has been abandoned, as has the traditional printing process.

The Working Notes series implies a format that is a work in progress that is cut-off at a set date and published. It will not be the definitive work; it will be as accurate as possible and will contain the latest thoughts on the various subjects covered.

Hopefully this format of book will be followed by several new books. Subjects such as raising fry, the keeping of seadragons, and a beginners book on 'how to' for the person new to the hobby are all in the pipeline.

New editions of this book will be published as and when important updates to the information we have provided is available.

As always, the most up to date information with the latest ideas on subject matter will be available on the Seahorse.org website.

Disclaimer

<u>Limit of liability & disclaimer of warranty</u>

The publisher and authors have used their best efforts in preparing this book. The publisher and authors make no representations or warranties with respect to the accuracy or completeness of the contents of this book, and specifically disclaim any implied warranties of merchantability of fitness for a particular purpose.

There are no warranties, which extend beyond the descriptions contained on this page. No warranty may be created or extended by sales representatives or sales materials.

The accuracy and completeness of information provided herein and the opinions and conclusions stated herein are not guaranteed or warranted to produce any particular results.

The advice and strategies contained herein may not be suitable for every individual. Neither the publisher nor authors shall be liable for any loss or damages, including but not limited to special, incidental, consequential or other damages.

Working Notes™ is a trademark of Creative Licence publishing.

Seahorse.org and Creative Licence publishing are not associated with any product or vendor mentioned in this book

18

Section 1: Anatomy

Seahorses are bony fish of the Family Syngnathidae and are in the Genus Hippocampus. They are characterized by having a bony body and a long, narrow, fused jaw. Among fish their external anatomy is unique.

External Anatomy

Orientation

This figure shows the areas of a seahorse and the terms used to describe them. The front view of a seahorse, looking into its face or belly, is the ventral orientation. The view from behind the seahorse is its dorsal orientation. The side view, used here in the illustrations, is referred to as the lateral orientation. The lower parts, or the lower section of any parts, are called the posterior, and the upper parts, the anterior.

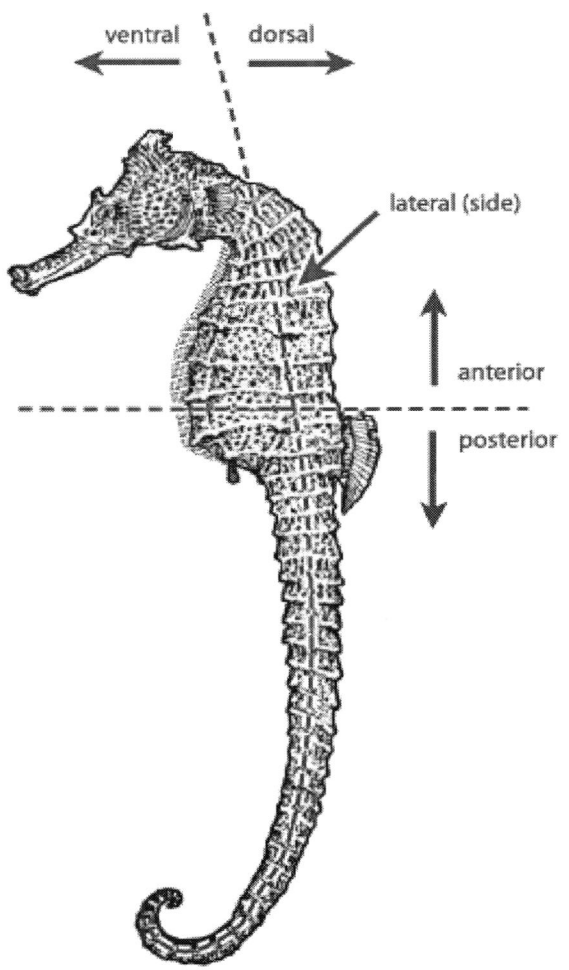

Sections and Measurements

The commonly described sections of a seahorse are the head, trunk and tail. The length of a specimen is measured from the top of the crown on the head of the seahorse to the tip of the tail.

The head is measured from the tip of the snout to just before the first trunk ring.

Snout measurement is made from the tip of the snout to the gill opening.

The depth of the chest is measured from the superior trunk ridge to the keel.

Seahorse length, snout to head ratio and chest depth are measurements that can be used to determine the species of seahorse.

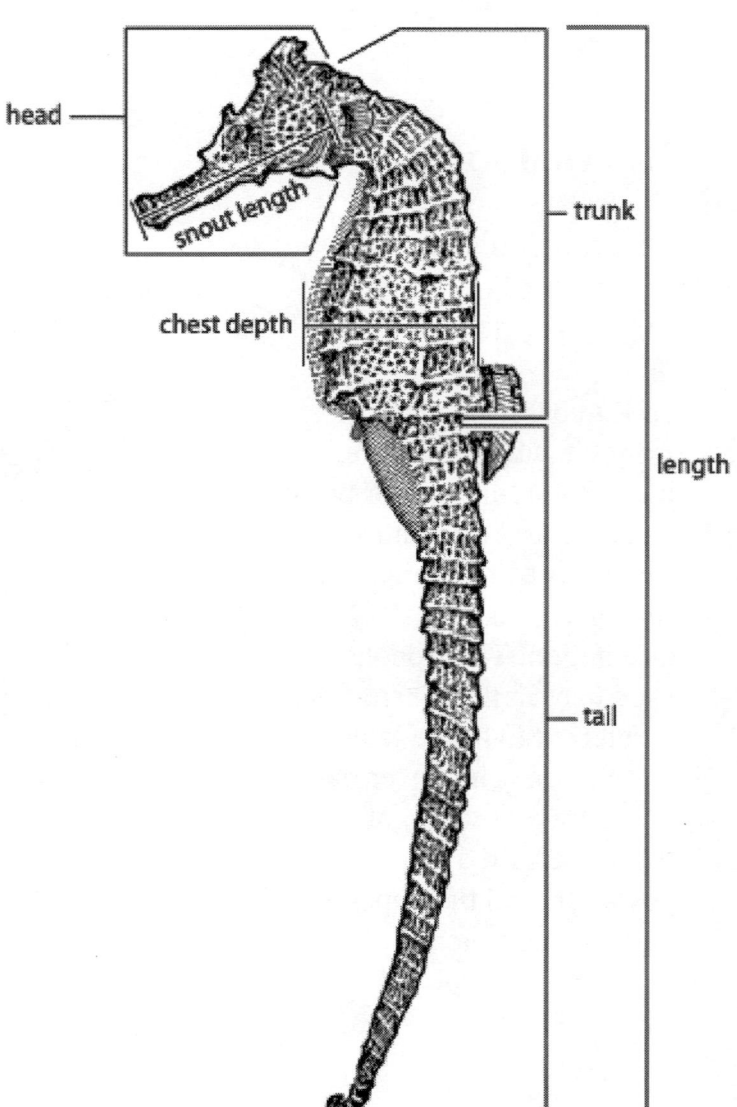

20

Parts of the Body

Trunk rings - one of the features used to determine the species of the seahorse - are counted from the first (the uppermost ring seen from the dorsal view) to the ring immediately above the anal fin.

Trunk ridges are the vertical spines running down the back of the seahorse from trunk to tip of tail (the superior trunk ridge), the spine running down each side of the seahorse trunk (the lateral trunk ridge), and the spine running along each side of the keel from neck to anal fin (the inferior trunk ridge).

Propulsion is by the pectoral fin, located just behind the gill opening, and the dorsal fin, which joins the trunk at the tail.

Tail rings are counted from the ring just below the anal fin to the ring before the tip of the tail.

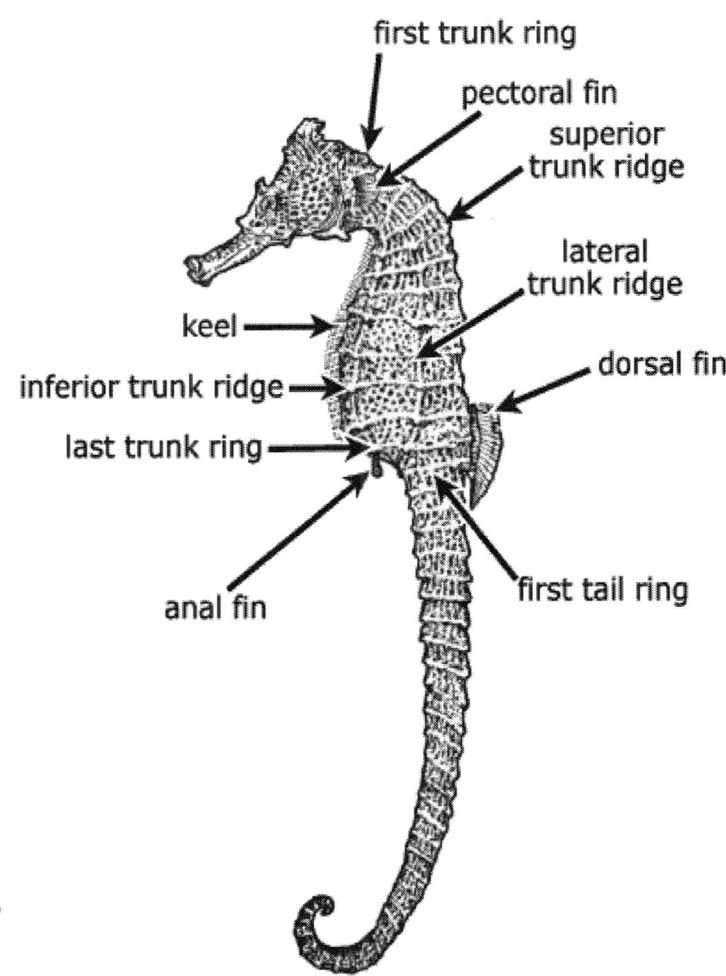

first trunk ring

pectoral fin

superior trunk ridge

lateral trunk ridge

keel

dorsal fin

inferior trunk ridge

last trunk ring

first tail ring

anal fin

Parts of the Head

Features of the head are illustrated on the right. The coronet can be low and fairly smooth on some species, to tall with pronounced points on others.

Eye, nose, and cheek spines also differ in length from species to species and also differ within a species from specimen to specimen.

All seahorses have independently moving eyes and a pair of pectoral fins immediately behind the gill opening.

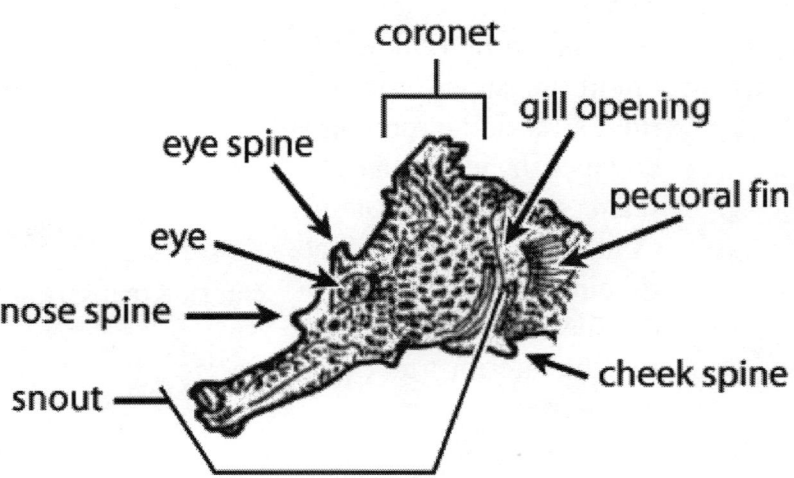

Some species of seahorses have spindly appendages, called cirri, in the area of the facial spines and trunk ridges.

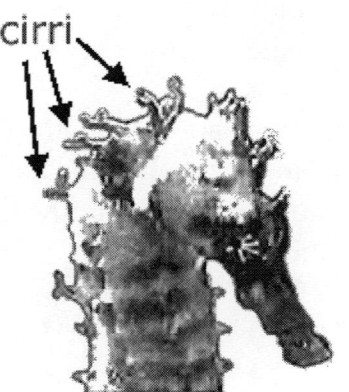

Sexual Characteristics

The external anatomy of female and male seahorses differs. This characteristic is called sexual dimorphism.

On the female seahorse, the lower abdomen joins the tail at a sharp angle and her anal fin is often higher and slightly larger.

On males, a brood pouch is found beneath the anal fin, and when empty, tapers gradually to the tail. During courting or when pregnant, the pouch is very pronounced and protruding. It features a vertical opening into which the female deposits her eggs, and from which fry emerge after gestation.

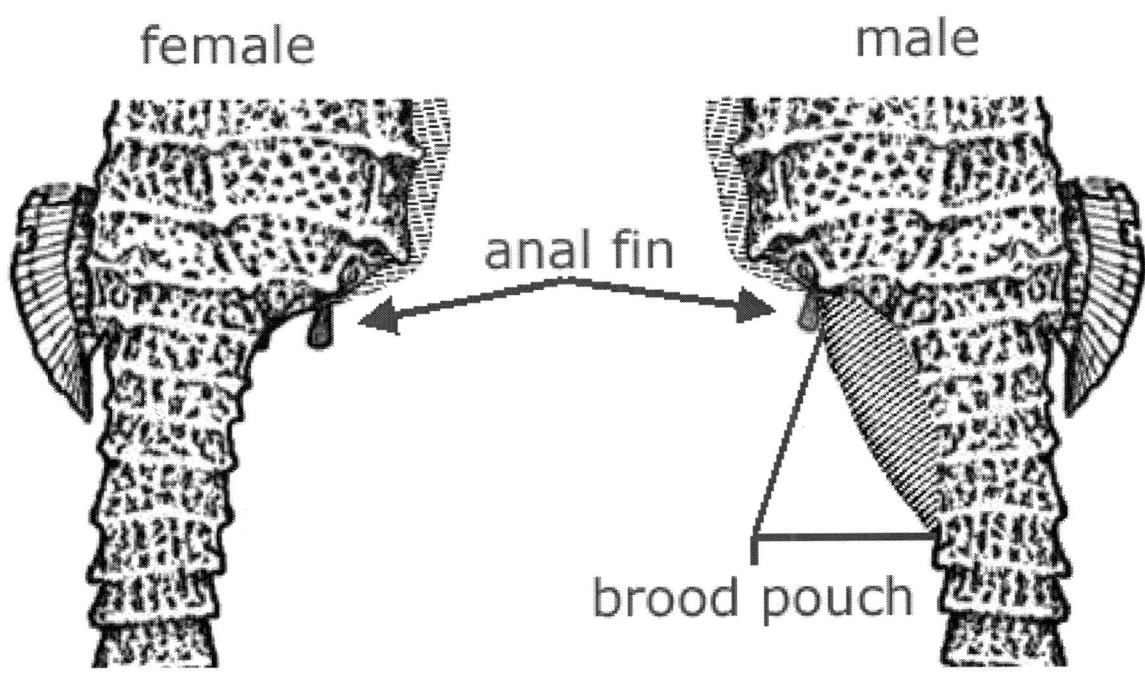

female

male

anal fin

brood pouch

Internal Anatomy

The fish inside

Male Seahorse Internals

The gills (1.) are located just behind the eye and appear as a red circular tufted area. Behind the gills, 2 structures can be identified, heart (2.) and esophagus (3.). The white tube-like esophagus is seen in the posterior position. Anterior to the esophagus is the pink heart. The yellow structure seen inferior to the heart and esophagus is the liver (8.). A second portion of the liver (8.) extends inferior to cover the intestines (7.). In the centre of the liver is the green gall bladder (4.) which contains bile. Between the gall bladder and liver is the translucent airfilled swim bladder (5.), which extends down to the bottom of the body cavity in some species. The sausage-like kidney runs along the anterior spine and extends from the level of the gallbladder down to almost the bottom of the visceral cavity. The most inferior tip of it is seen in this picture at the lowest posterior (6.) corner of the dissection.

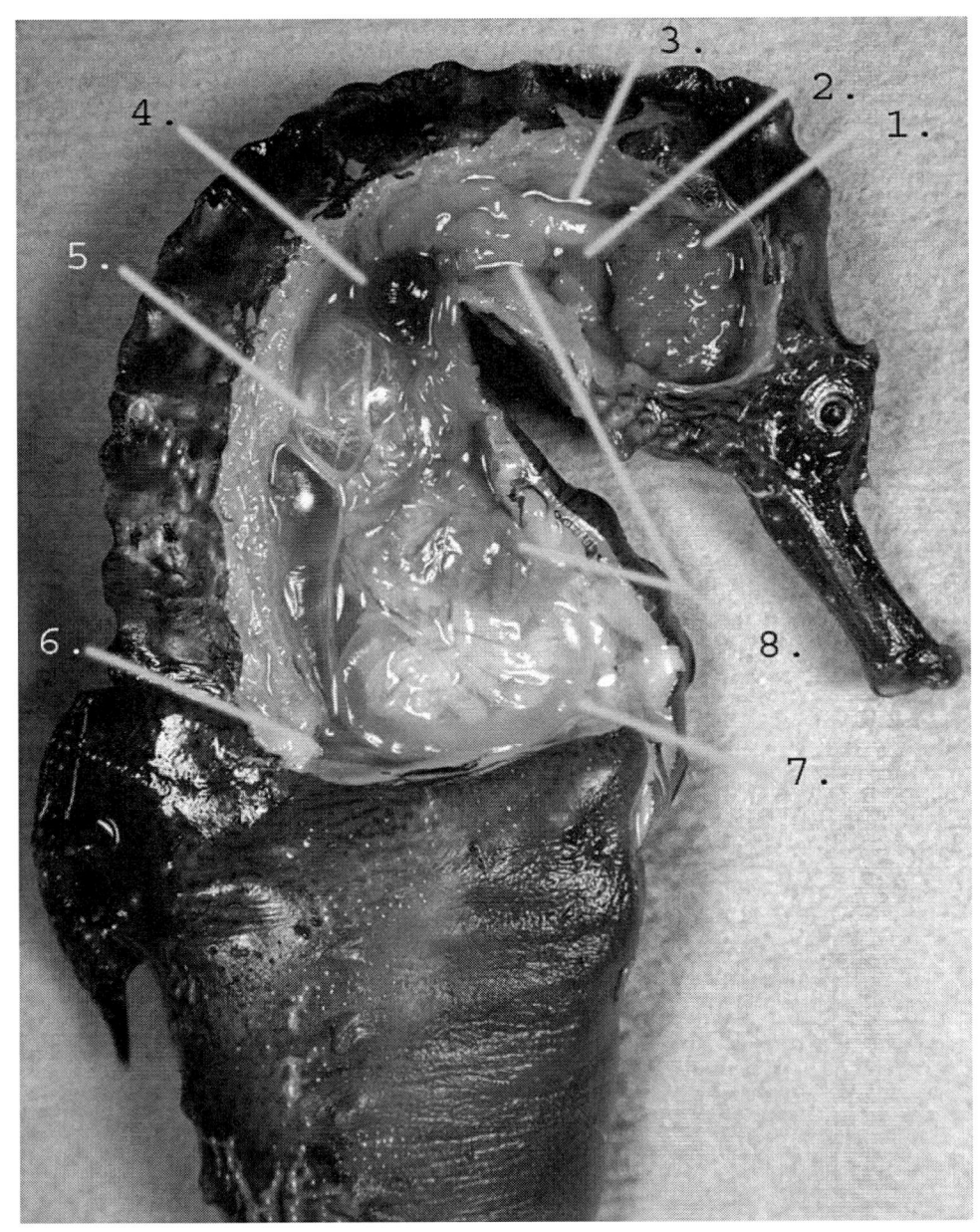

Female Seahorse Internals

The gills (1.) are located just behind the eye and appear as the red circular tufted area as in the male. There is a yellow structure seen inferior to the heart and esophagus which is the liver (8.). The intestines (7.) are clearly shown. Between the gall bladder and liver is the translucent airfilled swim bladder (5.), which extends down to the bottom of the body cavity in some species. This female has a full, healthy ovary (9.) which is egg laden.

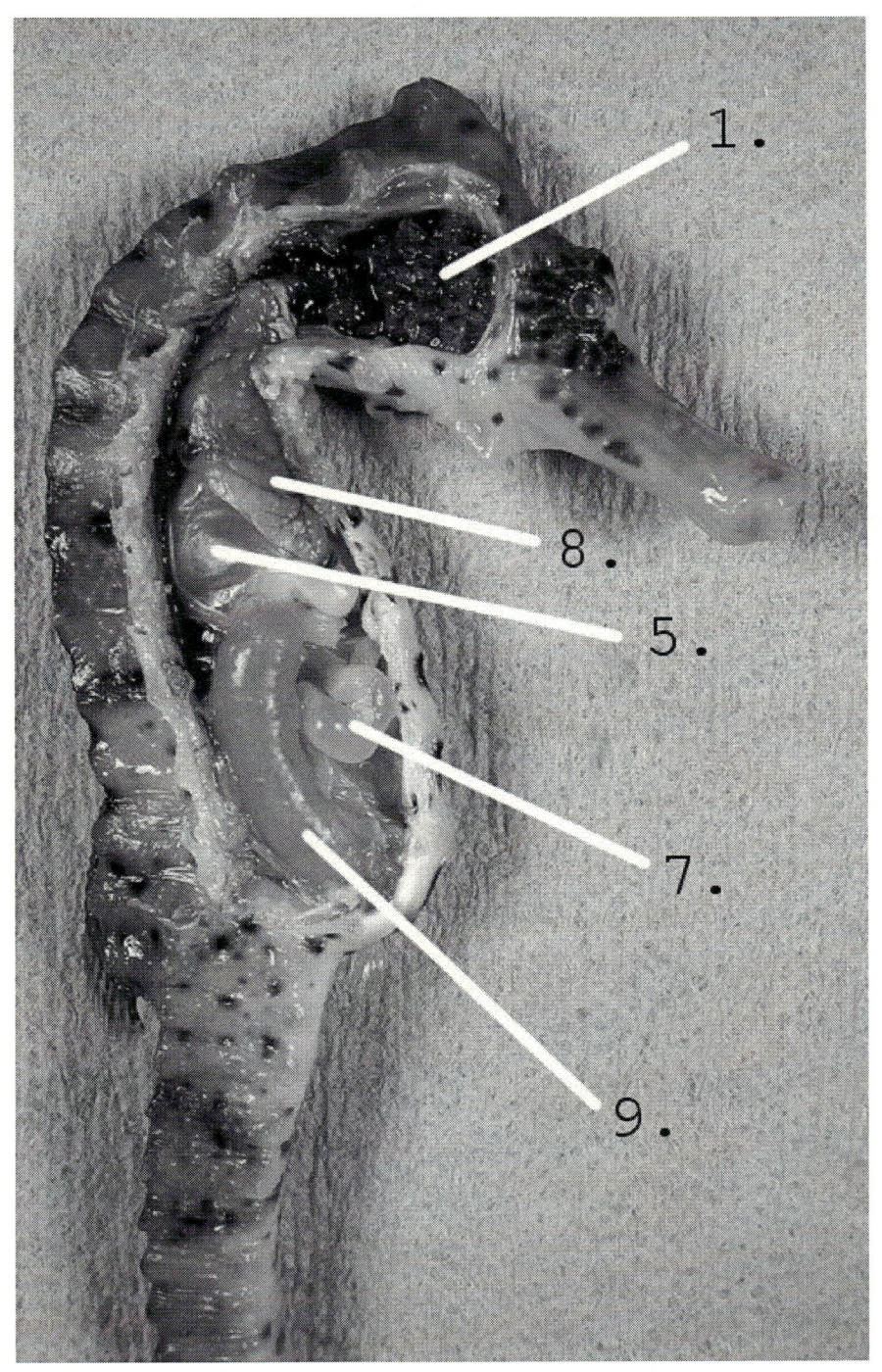

Internal organs

Internally the seahorse resembles most other teleost fish but due to its unusual shape the major organs appear crowded, especially in the neck region. The skin-covered snout is supported by bone and cartilage.

Masticatory muscles are used to activate a bone called the trigger, at the ventral base of the snout and in combination with movement of the opercula, and produce a powerful suction that partially liquefies the seahorse's prey. The vacuum chamber is created by a closed gill chamber which for the most part has a one-way valve system with respect to the gill openings. The gills are grapelike in appearance and also function as secretory organs giving off sodium and chloride ions.

Located immediately behind the gills is a two-chambered heart. It is fed by blood vessels bringing blood from the liver, kidney, tail and the rest of the body. The heart pumps blood to the gills where gas and ion exchange takes place. The oxygen rich blood from the gills is sent to the brain, heart, liver and kidney, with the rest of the body to follow. The tail is furthest from the heart and most likely receives blood with the lowest oxygen tension which may explain the common finding of tail lesions. However, the muscular action of the tail probably helps to propel blood back to the heart.

The brain is encased in a bony skull and the eyes are in bony sockets.

The liver is bi-lobed with half usually residing in the neck region and the other lobe in the anterior chest. This varies from seahorse to seahorse with some having both lobes in the chest. The gall bladder is usually found under the liver lobe that is most posterior. It is dark green and filled with bile.

The kidney is composed of two fused lobes that form a sausage shaped organ which is dark red to pink, and runs from the base of the neck to the bottom of the body cavity on the anterior surface of the spinal column. It is a type IV marine teleost kidney in that the superior and inferior poles are indiscernible from each other microscopically. In other fish one pole of the kidney has more bone marrow tissue than kidney tissue, while in seahorses both types of tissue are distributed equally throughout.

The seahorse kidney is also different in that it lacks glomeruli. Magnesium and sulphate ions are secreted in the proximal tubules which merge into a collecting system. Seahorses are hypo osmotic to their environment and must constantly drink seawater to stay hydrated. The gills and kidney function to maintain the ionic balance.

A true stomach is generally not appreciated in the seahorse and it also lacks a pyloric cecum. The GI tract is relatively short but does have numerous villi and intestinal folds lining the lumen to increase absorption. The GI tract is generally longer in the male of the species.

The air bladder is a gas filled organ used for buoyancy control that extends from the upper body cavity to the lower body cavity, tapering as it proceeds downward.

Sexual organs

Female seahorses have a paired right and left ovary which, when filled with ripe eggs, are orange in color. These connect with the ovipositor.

The picture of the female seahorse on the rear cover of this book clearly shows a healthy ovary, filled with eggs.

Males have testis which are paired and located as streak gonads on the surface of the kidney. They deliver sperm to the pouch.
The pouch is the gestational organ of the male. It is lined by a cuboidal epithelium which grows around incubating eggs. As such, it provides oxygen, nutrients and hormones to the eggs while maintaining a hypotonic environment with respect to the surrounding saltwater.

Section 2: Viral and Bacterial Diseases

Chapter 1 Viral

Introduction to viruses

Viruses are among the smallest of living things. They are composed of a protein shell containing genetic material which is either DNA or RNA, depending on the virus. The protein shell of the virus is adapted to attach to cell receptors on the host cell. Mutation of the shell proteins allows some viruses to adapt to new host species and evade attempts at vaccine creation.

Once attached to a living cell, a virus releases its genetic material into the machinery of the host cell where it hijacks resources to reproduce itself. Viral copies are then released to infect other cells either by rupture of the host cell or by budding. Viral diseases can be very contagious and strict quarantine of new arrivals is recommended to prevent their spread.

Lymphocystis Disease

Lymphocystis is a chronic, usually non-fatal, viral disease caused by an Iridovirus. Lymphocystis has been observed in most freshwater and saltwater species. Clinically, fish will present with variably sized white to yellow cauliflower growths on the skin, fins and occasionally the gills. Sometimes, in severe cases, the disease can be systemic with white nodules on the mesentery and peritoneum. The virus gains entry through epidermal abrasions where it infects dermal fibroblasts.

The histologic appearance is that of giant cells in the dermis, which can be up to 500X the size of a red blood cell. Also present is an enlarged nucleus with numerous cytoplasmic inclusion bodies. Inclusion bodies are viral proteins within the infected cell and are easily recognized under high magnification as pigmented globules that stain differently than the host tissue. The inflammatory response is variable. When present, it is most commonly a lymphocytic infiltrate.

Some believe the disease can lay latent only to emerge in times of stress or when poor water conditions are present. The disease is usually self-limiting and refractory to treatment. Nodules may last several months and cause affected fish to be susceptible to secondary bacterial infections, which should be treated accordingly. Attempts should be made to reduce stress and provide good environmental conditions. Removal of external lesions under anesthesia, followed by application of a topical medication is advocated by some, but this is best left to the professionals. Prophylactic antibiotic therapy is recommended after such procedures.

Experimental therapy with 6-mercaptopurine has shown some success. Once a fish is infected reinfection can always occur. Lymphocystis is the most common viral disease of aquarium fish.

However, based on the most current evidence, the disease is relatively uncommon in seahorses.

Symptoms

Variably sized white to yellow cauliflower growths on the skin fins and occasionally the gills
White nodules on the mesentery and peritoneum in severe cases

Treatment
Reduce stress
Improve water conditions
Treat for secondary bacterial infections as necessary

Chapter 2 – Bacteria

Introduction to bacteria:

Bacteria are present in almost every environmental niche on earth and are classified as prokaryotes, which means "before nucleus". This distinction was made to differentiate them from the eukaryotes, which means "with nucleus". Prokaryotes do not have a true nuclear membrane and are further classified into eubacteria and archaebacteria with the archaebacteria being thought to be a more primitive form. Archaebacteria, not considered to be of medical importance, are usually found in extreme conditions such as sulphur springs, hot springs and the Antarctic Ocean. On the other hand, Eubacteria constitute the bacterial forms that cause disease and also include cyanobacteria.

Bacteria possess no membrane bound intracellular organelles. Unlike eukaryotes, the machinery of the cell is found floating freely in the cytoplasm. All have a cell wall of some type and some have motility by means of flagella.

Methods to classify medically important bacteria have evolved in an effort to diagnose and treat the diseases they cause. Some of the first attempts included classification by size and shape as observed under the microscope, such as cocci versus bacilli (rods). In 1884 a Danish physician, Christian Gram, developed a stain that could divide bacteria into Gram positive or Gram negative groupings based on whether or not their cell wall retained the stain after washing with alcohol. Various other stains such as the acid fast stain were also developed to further divide and identify medically important bacteria.

Each bacteria has internal chemical processes and enzymes adapted to its environment. Some bacteria can ferment sugars. Some like low oxygen environments while others do well in high oxygen environments. Bacteria that thrive in high oxygen environments are called aerobic. Anaerobic bacteria prefer environments low in oxygen, or lacking oxygen altogether. Identification of these specific traits is also used to further classify bacteria into groups. Another tool, DNA analysis, has made recent advancements which have given scientists the ability to compare DNA of different bacteria and place them into groups according to their DNA homology. All of these methods of grouping can be used to identify bacteria in the laboratory and this facilitates a more reliable recommendation for treatment of bacterial infections.

Nocardia

Nocardia is a gram positive, aerobic, non-motile member of the actinomycetes family. Some strains are acid fast. In the lab it requires special media (Lowenstein-Jenson) and takes a long time to grow (21 days). Diagnosis is most easily made via gram stained material taken from lesions where the characteristic growth pattern of right angle branching can be observed. Nocardia is commonly found in soil and can affect both freshwater and saltwater fish. Clinical signs and symptoms are very similar to those produced by Mycobacterium (another bacterial disease discussed in full later in this chapter). The organism can be found in white punctate granulomatous lesions on the skin surface, in the muscles, or associated with the viscera.

The microscopic appearance shows either granulomata or abscesses containing right-angled branching bacteria best viewed with Gram stain or Acid Fast Bacillus Stain. Detection by immunoflourescent techniques also exists but requires specialized equipment.

Symptoms

Small punctate creamy-white lesions on the skin surface
Anorexia
Listlessness
Swollen abdomen

Treatment

If a diagnosis of nocardia is made or suspected, immediate isolation of the affected seahorse is critical to keep the infection from spreading. At this time there is no proven effective treatment for Nocardia.

Experimentation with sulphonamides has reported some success. When access to diagnostic equipment is not available the conservative course of action is to treat for mycobacterium since the clinical presentation is so similar to Nocardia.

In confirmed cases of Nocardia it may be prudent to euthanize the affected individuals unless experimental humane drug therapy is to take place in an effort to find a cure for this disease. All normal appearing tank mates must be closely observed for signs and symptoms and isolated as lesions appear. In cases of epidemic it is advisable to tear down and sterilize the tank, equipment and contents before restocking since the organism can survive for long periods of time in a tank environment.

Attempts at vaccine creation have been done on an experimental basis but no such program is known to exist at this time.

Mycobacteria marinum

Tubercular disease in saltwater fish is caused by *Mycobacteria marinum*, an acid fast, non-motile, gram-positive rod. The disease has worldwide distribution. Freshwater varieties exist and are encountered in farmed fish. In general, the disease shows slow progression with granuloma formation in the affected sites which can include muscle, skin and the internal organs. Skeletal involvement can result in deformities. Organ involvement can lead to death. Transmission is believed to be through ingestion. It can resemble Nocardia clinically.

The microscopic appearance shows granulomata containing acid fast bacilli. The bacteria require special media (Lowenstein-Jenson) to grow in the lab and are slow growing. Fluorescent techniques for diagnosis exist but require specialized equipment.

Symptoms

Lethargy
anorexia
emaciation (wasting)
nodular growths
skeletal deformities

Treatment

Mycobacteria found in fish are usually resistant to anti-microbial therapy. Resistance to Isoniazid is widespread. Multidrug therapy such as doxycycline and rifampin is currently the treatment of choice but is out of the hands of most hobbyists. The use of a UV sterilizer can help cut down on spread of the disease. Currently there is no vaccine available.

Vibrio

Vibrio is a common waterborne motile gram-negative bacterium. It has been linked to outbreaks of disease in both human and wildlife populations. Many different species of Vibrio exist and all are considered pathogenic to man. In aquarium fish, including seahorses, this pathogen can also cause serious disease and death. However, it appears that either non-virulent strains exist, or some fish may be asymptomatic carriers since Vibrio has been cultured from apparently healthy specimens. Vibrio can be divided into halophilic and non-halophilic species depending on the proclivity to grow in saltwater or not. It is easily cultured in the lab but does not grow well at temperatures below 18 C/64 F.

The different strains of Vibrio can show various patterns of disease which are also influenced by the host species. In general, there are types like *V. vulnificus*, which tends to be a more ulcerative and aggressive disease than *V. alginolyticus*, which attacks organ systems and can present as renal insufficiency (bloating, dropsy) or liver failure (sluggishness, seizures, anorexia). The seahorse tail is a common site for pathology. Many diseases commonly occur or present in the tail including Vibrio, Nocardia, Glugea, and GBD (known as "gas bubble disease" and discussed in full in Section 4). The lesions of Vibrio tend to present as erosion of the flesh or tail/snout rot. Discrete pus filled abscesses can also occur. This variability of symptoms makes diagnosis difficult. Since the bacteria grow best at higher temperatures outbreaks have also been associated with high tank temperatures, especially in the late spring and summer.

Vibrio species constitute the majority of bacterial disease in seahorses so empirical treatment for Vibrio is recommended whenever any bacterial disease is suspected.

Symptoms

erosion/sloughing of the flesh
cloudy eyes
rapid breathing
generalized swelling (especially with negative buoyancy)
swelling about the eyes and snout
seizures
anorexia
sluggishness
tail rot/snout rot

Treatment

Treatment is based on antibiotic susceptibility data obtained from over 40 Vibrio isolates taken from seahorses. In almost all isolates susceptibility to Ciprofloxin, aminoglycosides (such as kanamycin, gentamycin, neomycin etc.), and combination drugs such as Bactrim was found. Treatment experience has been mostly with temperate species like erectus and reidi but the same principles could be applied to other species. At this time the combination of lowered temperature (~12 C/70 F), an aminoglycoside, and a sulpha drug is the treatment of choice. Dosing should take place in a quarantine tank with special attention being paid to water quality. Temperature should not be lowered too quickly and the lights dimmed to reduce stress. Biofiltration is unnecessary as the drugs used are biofilter toxic. Water quality must be maintained with water changes made with well aerated saltwater. As with any disease state, nutrition becomes a key element for recovery. It is imperative to keep the seahorse eating or encourage it to eat with live food, if available.

Gut loading or enriching with Beta glucan and vitamin supplements is beneficial and recommended. Application of topical antiseptics to open sores or wounds should be performed as tolerated every 8 hours. This should be done in a separate bowl of tank water so as to not spill excess antiseptic into the tank.

Some strains of Vibrio can cause a kidney infection with renal failure. These animals will have increased body fluid and appear swollen. The edema results in increased mass which makes the seahorse sink rather than float as it might with GBD. While this makes it easier to tell the two conditions apart, the distinction is not that important with respect to Vibrio therapy. The standard treatment of choice for GBD is a drug called Diamox (acetazolamide) which has a diuretic effect that can help alleviate the swelling seen with some Vibrio infections. To date there is no evidence of any drug reactions while using Diamox in combination with antibiotics.

Commercial vaccines to coldwater strains of Vibrio encountered in salmon and other farmed fish exist. Cross reactivity of these vaccines with strains found in seahorses has not been investigated. A polyvalent vaccine to strains seen in seahorses would be a great boon to the hobby, but still presents significant problems.

Experience with commercial vaccines show that the mode of vaccination has an impact on how effective the vaccine is in providing immunity and the length of that immune state. Injection is most effective and long lasting, followed by ingestion, and finally external exposure (dip). Due to seahorse anatomy and diet, both injection and ingestion are not without difficulties.

It is Seahorse.org's understanding that an experimental vaccine was used as a dip on seahorses at Shedd Aquarium in Chicago but that a live Vibrio challenge was never performed.

The effectiveness of such a vaccine in seahorses has yet to be proven.

Seahorse.org is working on a vaccine specifically aimed at the common Vibrio strains found in seahorses.

Section 3 – Parasites.

Chapter 1 – Worms

Introduction to Worms

Worms of various species that infect fish are found throughout nature and some are transmittable to man such as *Diphylobothrium latum*. Classification of worms is the result of comparative anatomy studies and similarity of life cycles. Except for monogenetic trematodes, most parasitic worms require at least one intermediate host to successfully spread. Intermediate hosts include various crustacea and mollusks. Birds, mammals, fish, amphibians and reptiles can be the definitive host species affected. The life cycles of these parasites can be complex, sometimes involving more than one intermediate host. Specificity to certain hosts and intermediate hosts does occur but varies from species to species.

Infestation by worms is common in wild caught specimens. In a series of 43 wild caught and captive bred seahorses, 18 (42%) were reported to have worms at autopsy and in 6 *H. erectus* and 1 *H. reidi* (16%) it was the cause of death or a major contributor. Quarantine and deworming are essential to successful husbandry of wild caught species. These diseases tend to be generally well tolerated by the host, presenting as a chronic wasting condition. However, they do take their toll over time. For monogenetic trematodes dissemination becomes an issue for captive bred seahorses exposed to them. Worms requiring intermediate hosts are thought not to reproduce in the home aquarium. However, infestation via introduction of an intermediate host containing wild caught food items (e.g., crustacea) is possible.

Infestation by larval worms can be more damaging than those by adult worms since the worms can migrate through tissue and organs leaving destruction in their wake. Adhesions between organs and damage to viscera and muscle can be extensive. Some types can affect the eye causing blindness. Virtually any organ system can be involved. Identification of larvae can be done on affected tissue at low magnification. In wet preparations movement can be identified. Egg masses replacing the gas bladder have been identified in hippocampus species. Exact classification of the causative agent has not been performed but it is presumed to be a trematode. Masses of this kind, especially involving the swim bladder, can cause problems with buoyancy. Through mass effect other organs may be displaced and suffer damage.

Cestodes (tapeworms)

Tapeworms are segmented worms that occupy the gut lumen as an adult but can also occur in the larval form in muscle, organs, or in the body cavity. Adult worms (Eucestoda) are characterized by having an attachment organ (scolex) armed with tiny hooks, suckers, and/or spines by which it feeds on nutrients taken from the gut wall of the host species. In some species (Cestodaria) the adult lacks a scolex, but this is less common and occurs mostly in primitive fish such as sturgeon. Being segmented worms, tapeworms grow by adding segments called proglottids. Proglottids increase in size along the length of the worm's body and eventually fall off.

Adult tapeworms have both pairs of sexual organs (hermaphrodites) and can fertilize their own eggs. Egg filled proglottids or eggs pass out of the host with feces to procreate the species. Cestodes often have complex life cycles requiring more than one intermediate host. Adult tapeworms tend to be larger than most other parasitic worms encountered and can reach several feet in length. Low power examination of feces for proglottids or eggs either as a wet prep or in histologic sections, is the easiest way to make the diagnosis. Most worms are so large as to be identified by the naked eye. Classification is done by characteristics of the scolex and body segments. Larval infestations can be very destructive. In autopsy series adult tapeworms were not identified in captive seahorse populations.

<u>Symptoms</u>

chronic wasting
worm protruding from anal opening.
ulcerations, sores, abdominal distension, blindness and organ
damage from migrating larval forms have been identified.

<u>Treatment</u>

Proper quarantine of wild caught specimens and prophylaxis with
praziquantel.
Treat for secondary bacterial infections as necessary.

Trematodes (flukes)

Trematodes are divided into either monogenetic or digenetic varieties depending on the need for a secondary host. They are generally small in size (1 to 5 mm) and can occur at various sites. Autopsy results show that they are fairly common in wild caught seahorse populations where they are found at various sites in the gastro-intestinal tract and in the gills.
Both larval and adult trematodes can occur in fish with subsequent disease states. Due to their direct life cycle it's theoretically possible for monogeneans to proliferate in the home aquarium. The complicated life cycles of digeneans are less likely to support such spread but it is possible in large systems with many species of potential intermediate hosts.

Diagnosis can be made by identification of adults or juveniles in wet preparations taken from external lesions or through examination of tissue sections. Eggs may be seen in fecal matter of the infected host.

Speciation is not critical since therapy is similar for all worms. Besides the damage done by migrating juvenile forms, secondary bacterial infection can pose a problem.

Monogenetic trematodes have a direct life cycle and do not require a secondary host. Most are species and site specific, mostly oviparous, releasing a small number of eggs into the environment via the feces of the host or directly into the water in external infestations. Free-swimming ciliated larvae, called oncomiracidium, hatch from the host. As such, these larvae must find a suitable host within hours - before perishing - to complete the cycle. Upon finding a suitable host the oncomiracidium migrate to their final attachment site and mature into adults where it is thought they remain for life.

Most occur as external infections involving the skin, gills, body, and fins. Less common species have been reported to infect the oral cavity, nasal cavity, intestines, ovipositor and urinary bladder of fish. Monogeneans are further divided according to the morphology of the opisthaptor (external attachment organ) which can be a combination of suckers, clamps and hooks.

Digenetic trematodes constitute a very large group of internal parasites. Having very complex life cycles, they require one or more intermediate hosts. All are oviparous, releasing eggs through the host feces which hatch into a small ciliated form called a miracidium. The miracidium has a few hours to find the first intermediate host which is almost always a mollusc or gastropod. These can then develop into free swimming cercaria or metacercaria encysted in tissue.

Eventually the final host and site is reached where the adult form develops. Digeneans are characterized by the morphology of the ventral holdfast organ and oral sucker.

They differ from monogeneans in not having chitinous hooks. Adults are generally found in the intestines and gastrointestinal tract but have also been seen in the swim bladder, ovary, body cavity, urinary bladder and in the circulatory system of fish. Adult worms in small numbers are usually not a serious problem but do present a drain on general condition. Migrating larva can cause considerable internal damage including blindness and death.

Symptoms

chronic wasting
ulcerations, sores, abdominal distension, blindness and organ
damage from migrating larval forms have been identified.

Treatment
Proper quarantine of wild caught specimens and prophylaxis with
praziquantel. Treat for secondary bacterial infections as necessary.

Nematodes (roundworms)

Nematodes are segmented round worms that can infect fish as either a secondary or final host. Adults are almost always found in the digestive tract where they release live young or eggs which hatch into a free-swimming larva.

Larva are usually eaten by a crustacean which is then eaten by a fish. The larva can then develop into an adult if the fish is a final host, or encysted where it waits to be devoured by the final host which can be a fish, mammal, bird or other large predator. In captive bred fish populations, nematodes are generally not a problem since proper intermediate hosts are not present. Adult worms can be found in wild caught specimens. Like many other worms, adults in small numbers are tolerated well. Migrating juveniles appear to cause more harm and promote secondary bacterial infections.

Nematodes are best identified in wet preparations and tissue sections. They do not have hooks or suckers and are classified by size, the morphology of the head, esophagus, tail and excretory pore. Adult worms can vary in size from a few mm to 10 cm and greater in large animals. Eggs may be seen in fecal matter of the infected host.

Symptoms

chronic wasting
worm protruding from anal opening.
ulcerations, sores, abdominal distension and organ damage from migrating larval forms have been identified.

Treatment

Many antihelminthics (dewormers) are effective against adult nematodes. They tend to work best when administered orally. However, since wild caught specimens may harbour a variety of parasites which are most often not diagnosed it is best to treat with praziquantel. No drug has been proven to eradicate encysted larva. Treat for secondary bacterial infections as necessary.

Acanthocephala (spiny-headed worms)

Acanthocephalans are generally small worms (mm to cm) that live in the intestinal tract where they use their spiny head to anchor in the wall of the bowel. They have no mouthparts as nutrients are absorbed directly through their surface from passing food items. Their life cycle is similar to nematodes except there is no free-swimming larval stage. Eggs must be eaten by the first intermediate host which is usually a crustacean. They also share the same features with nematodes in regard to pattern of infestation and the resulting morbidity and mortality. Studies have shown they are not prevalent in tropical marine reef fish.

Acanthocephalans are best identified in wet preparations and tissue sections. They are classified by size, the morphology, and arrangement of hooks on the proboscis. Eggs may be seen in fecal matter of the infected host.

Symptoms

chronic wasting
Abdominal distension, and organ damage from migrating larval forms have been identified.

Treatment

Treatment is similar to that described for nematodes.

Chapter 2 – Non worm diseases

Glugea

Glugea is a member of the microsporidea family and is an obligate intracellular parasite that spreads via unicellular spores. Microsporidea have been found in both fresh and saltwater fish as well as numerous land animals. Historically they have been difficult to classify and through DNA comparison studies are now thought to be a member of the fungus family. Microsporidea also includes Heterosporis, Microgemma, Pleistophora, and Nosema. All have small (microns) microspores at some point in the reproductive cycle but Glugea seems to be the most important with respect to Hippocampus species.

Glugea lesions, called xenomas, can occur in the muscle, dermal skin, liver, bile duct, kidney, ovary, central nervous system, GI tract, peritoneum, gills, and even the eye of various species. The mature cystic xenomas are relatively large (mm to cm) and look similar to clusters of white grapes. The lesions contain numerous microspores which spread the disease after the xenoma matures and ruptures. Infection occurs by ingestion of the spores.

Autoinfection from lesions in the same host is thought to occur but has not been proven. While low temps (below 15- 18 C or 59- 64 F) slow the growth of the organism, spores have been proven to remain viable in water at 4 C/39 F for at least one year. There is no known method to culture the organism in the laboratory (in-vitro) since it requires a living cell to multiply.

Some species of microsporidea can affect a wide range of host species while others are more specific. The pathogen infects living cells, transforms them into spore factories, and is always detrimental to the host. Some infestations appear to take longer to develop while others can be more aggressive, killing the host in a short period of time. *H erectus* has been associated with *Glugea heraldi* and experience with the disease has shown it can manifest as a chronic condition over a long period of time. Eventually the seahorse will succumb due to local tissue destruction, abdominal distension, and the loss of essential nutrients to the developing xenomas.

Diagnosis is most easily made by biopsy where the characteristic appearance of the minute microspores is diagnostic at high magnification (1000x).

Symptoms

Small white grape like lesions on the body, fins and tail. Can also occur as single cysts.
Anorexia
listlessness
Abdominal swelling

Treatment

At this time there is no known treatment for microsporidean infections. The similarities to fungus have prompted the use of experimental anti-fungal therapy with little or no success. Fenbendazole which blocks microtubule formation has also been tried and does not appear to alter the course of the disease.

Since Glugea has signs and symptoms that are similar to other treatable diseases, strict isolation and a course of antibiotic therapy is warranted whenever a tissue diagnosis is not available. No known attempt at vaccine production has been tried. Once a definite diagnosis of Glugea has been made the best plan of action is to humanely euthanize the affected individuals and to sterilize the tank, its contents, and all equipment with bleach.

Invertebrates, macro algae and live rock are not subject to the disease but potentially can never be certified pathogen free and will carry the risk of accidental spread to other tanks via fomites. Exposure to unaffected seahorse populations even after lengthy tank quarantine is highly discouraged. Unfortunately, extreme action is warranted involving sterilization of all such material.

Protozoans

Protozoan Infections. Marine Ich (*Cryptocaryon irritans*), Marine Velvet (Amyloodinum, *Oodinium ocellatum*) and Uronema

Protozoans are eukaryotes that have a well-defined nucleus and cytoplasmic organelles. They are larger and more complex than bacteria. For the most part they are not commonly seen in seahorses but cases do occur from time to time. While uronema, which can cause ulcerations, can directly affect the skin of the seahorse, Ich and Velvet are more commonly seen in the gills or on the fins where they present as white spots. They can be difficult to speciate by the untrained eye, but just the presence of a unicellular organism, especially cilliated ones, can be helpful in ruling out primary bacterial infections which require different treatment.

Marine Velvet Disease

In "Marine Velvet Disease" (Amyloodinum, *Oodinium ocellatum*), the motile infective stage is the dinospore. It can remain viable at specific gravities from 3 to 45 ppt. and temps from 7-30 C/44-86 F. The organism is more complicated than *Cryptocaryon irritans*. Achievable temps and salinity are not an effective weapon against it. Copper seems to put some of the non-motile tomonts in a state of non-division. When copper is removed some start to divide again. Dosing the tank with the fish in it often results in reinfection through this process of reactivation. Formalin (6 to 9 hours) can detach some trophonts from a fish but some may resume activity once the formalin is removed. So treatment must ensure that enough time is given to allow all forms to develop into vulnerable motile dinospores.

Velvet primarily attacks the gills and can be quick and deadly in closed environments or heavy infestations.

Treatment should be done in a quarantine tank and the tank laid fallow (fishless) for 6 weeks to insure eradication of the organisms. Ultra violet sterilizers can help cut down on disease transmission.

Symptoms

increased respiration
scratching
white spots on fins and eyes
anorexia

Treatment

Copper (3 weeks)
Formalin
Eradicating this parasite from the infected tank may require laying the tank "fallow" for a period of 6 weeks.

Marine Ich

Marine Ich (*Cryptocaryon irritans*) is a ciliated protozoan that is an obligate ectoparasite which means that it is an external parasite that needs the fish host to complete its life cycle. *C. irritans* has a four stage life cycle, including the parasitic stage (theronts), which produces the appearance of white spots all over the fish (3 to 5 days).

These are small when they first attach, but once mature, they drop off the fish where they encyst and begin to reproduce. This is called the tomont or reproductive stage. The tomonts divide for a number of days (3 to 28 days) after which the cyst ruptures, releasing the tomites.

Tomites may differentiate into theronts, which actively seek a host to reinfect (24 to 48 hours). Affected fish develop white spots on their skin, gills, fins, and eyes. In seahorses, the skin is usually not affected. Symptoms are similar to Velvet.

Treatments include copper in a quarantine tank, and hyposalinity. Six weeks of hyposalinity is thought to eradicate the organism from the display tank. Fresh water dips have not proven to be effective. Researchers have also found the use of cleaner wrasse (*Labroides dimidiatus*) not to be effective against the disease.

Symptoms

increased respiration
scratching
white spots on fins (~.5 mm)
anorexia

Treatment

copper
hyposalinity

Uronema

Uronema is a ciliate unicellular parasite that can cause serious skin, gill, and internal lesions in fish over a wide range of temperatures and salinity. Along with skin ulcers, involvement of the kidney and liver can occur. The organism is egg-like in shape, surrounded by short cilia, with one long caudal cilium, and measures 50 to 100 um in length. This disease has been reported to affect seahorses and other syngnathids. The best defense against this disease and all other protozoan infections is to use strict quarantine for all new arrivals. Hyposalinity is ineffective in treating uronema.

Since it is sometimes difficult to differentiate one ciliate from another, some experts advocate treating them all as uronema with formalin. A combined formalin/hyposalinity treatment has also been advocated. Symptoms common in seahorses include increased respiration, pale discoloration and loss of skin with ulceration. It is important to attempt to identify the organism on wet prep when skin ulcerations are present to differentiate it from Vibriosis.

Symptoms

Scratching
Pale discolorations
Weight loss
Rapid breathing
Skin Lesions
Ulcerations

Treatment
Fresh water dip followed by prolonged formalin immersion

Parasitic Copepods

Copepods are crustaceans that are found in both fresh and saltwater as free living and parasitic forms. Free-living forms (greater than 10,000 species) tend to be small, ranging in size from less than 1mm to a few millimetres in length. Parasitic forms (approximately 1700 species) can become quite large. Most copepods encountered in the home aquarium are beneficial non-parasitic benthic detrivores.

All are characterized by having a single simple eye and complex life cycles. Females carry eggs which release a free-swimming nauplii. In most parasitic species only the female is parasitic while the male remains free living or parasitic on the female. After a number of successive moults, nauplii become preadults and finally adults. Adults attach to their hosts and feed through the use of piercing and sucking mouthparts. Generally they are found on the skin or in the gill cavity. Endoparasitic varieties exist but are uncommon.

The two most common types of parasitic copepods are from the families Ergasilidae and Lernaeidae, with Ergasilidae being more common. The species can be classified through the morphology of their buccal region.

Ergasilidaens tend to retain the overall morphology of free-living copepods, while Lernaeidae are more grub-like. Both are most commonly found infecting the gills and oral cavity.

Caligariform types (sea lice) infest the skin and have been known to cause serious problems in cultured fish, including salmon. Morphologically they resemble Argulus (fish lice) not only in appearance but also in common name.

Caligariform types can multiply quickly in aquaria and closed environments.

Leraeid types (anchor worms) possess anchor-like processes to attach themselves to the host. They resemble leeches but do not have posterior suckers.

The problem is most common in wild caught species that have not undergone proper quarantine where they would have been identified and treated. It can also occur in any species exposed to infected individuals or parasites brought in with live rock or invertebrates.

Parasitic Isopods

Five families of isopods contain species that are parasitic on fish. Some are parasitic during part of their life cycle (Gnatthiidae, Cymothoidae) while others (Anilocridae) are permanent obligatory parasites. They are not commonly seen in the home aquarium and are more common in nature. If encountered, symptoms and treatment are similar to those given for copepods.

Damage is caused to fish by irritation of the skin and gills at attachment sites. Infestations are usually not serious, except in small fish or in a closed environment. Affected individuals will show signs of scratching or erratic behaviour. Severely affected individuals can have rapid respiration and failure to thrive. Secondary bacterial infections can occur. Parasites can be identified with the naked eye or under magnification depending on their size. Although not useful for eradication, sometimes a fresh water or formalin dip can dislodge parasites for closer examination and diagnosis.

Symptoms

Visible parasites or attachment sites
unexplained sores or wounds
Scratching
Increased respiration and failure to thrive in severe cases

Treatment

Anchor worms are best and most easily treated with removal under anesthesia followed by a topical antiseptic.

Sea lice present more of a problem and removal can usually be accomplished with organophosphates but resistance has been reported. Copepods are usually not susceptible to copper or formalin therapy. Some have reported relief with the use of parasite eating wrasses. Ivermectin is effective but seems to be not well tolerated.

Tank infestation may be a problem with parasitic copepods. Organophosphates will also eradicate the organism but are not recommended for use in the display tank, with invertebrates, or live rock. Some advocate a total of 3 treatments at 7-day intervals to entirely eradicate the organism. The required fish free tank fallow time to safely eradicate these organisms is not known at this time.

In all cases surveillance and treatment of secondary infections is warranted.

Fish lice

Fish lice are crustaceans of the class Brachiura. They are among the largest fish parasites and can cause considerable damage to their hosts. All are parasitic on the skin of fishes (occasionally amphibians) but may also invade the bronchial cavity (gill cavity).

These parasites are dorsoventrally flattened, with a wide shield over their thorax. The eyes are paired, sessile and compound. The first antennae and second maxillae have claws for gripping the host. These are used to attach to the host and move across the host's body. The mouth is cone-shaped and equipped with a pre-oral sting. The sting is said to inject a toxin produced by a gland at its base that may be responsible for tissue destruction at the attachment site. Through the use of sucking mouthparts, the parasite feeds on host blood and interstitial fluids.

Fish lice have a direct life cycle and require no intermediate host. Therefore they can reproduce in the home aquarium. These parasites can swim from host to host and drop free of the host to lay their eggs. The eggs are laid in strips, glued to plants and stones. The young crustaceans hatch after about 4 weeks. After hatching a suitable host is found and after 5 to 6 weeks sexual maturity is reached. Most lice including Argulus cannot survive for periods of greater then 3 weeks without a host.

Most are less than 5mm in length but can overwhelm a small fish, particularly if present in large numbers. Fish lice have been considered to be involved in the spread of lymphocystis, ichthyophonus and to open the way for a secondary microbial infection.

More dangerous for the fish than the loss of nutrients to the louse is the possibility of a secondary infection, which occasionally arises after the infestation.

Some believe fish lice to be responsible for the spreading of infectious abdominal dropsy of cyprinids. Fungi may settle in the injection wounds. Also, the blood parasites Cryptobia are most probably transmitted by argulids. There does not seem to be any fish species which is not affected by this parasite.

Seahorses are usually affected by Argulus fish lice, which measure from a few mm to about 1 cm in size. Affected specimens can exhibit erratic behaviour as they try to remove, and react to, the irritation of the parasitic attachment. The characteristic "manta ray" appearance of Argulus is diagnostic.

Anemia may result from infestation. Increased respiration and failure to thrive may also be clinical features of the disease. Abandoned attachment sites are characteristically rounded and may show small puncture marks. Since they can become infected, ulcers and sores are also a characteristic of the disease.

Fish lice are best observed through low power magnification although they can be seen with the naked eye. Low-level infestations represent no significant direct threat to otherwise healthy adult seahorses. Heavy infestation can lead to serious illness and death.

Symptoms

Visible parasites or attachment sites
unexplained sores or wounds / scratching
Increased respiration and failure to thrive in severe cases

Treatment

Treatment is parasite removal under anesthesia followed by application of topical antiseptic. Secondary infections and transmitted disease are treated accordingly. Lice can also be removed with organophosphates or by prolonged formalin immersion when mechanical removal is not technically possible.
(1 : 4,000 for 1 hour).

If tank infestation is suspected, all potential host species must be removed and the tank lay fallow for at least 8 weeks. Formalin and organophosphates will also eradicate the organism but are not recommended for use in display tanks, with invertebrates, or live rock.

Leeches

Leeches are parasitic worms that feed off the blood of their hosts. They have a segmented body and distinct anterior and posterior suckers. They vary in size from a few millimeters to many centimeters in length. As parasites of fish they are found attached to the skin surface, gills or oral cavity.

Leeches have a direct life cycle. The young hatch from cocoons produced by hermaphroditic adults. Like fish lice, infestations are usually only serious when present in large numbers or through transmission of disease and secondary infections at attachment sites. Leeches have been suspected to transmit protozoan disease including trypanosomes and Cryptobia.
Through the use of their oral and posterior suckers they attach at two points to the host and extract blood and interstitial juices. A potent anticoagulant produced by the leech keeps blood flowing and its action has proved useful in medicine. Leeches can be found in captive seahorse populations, mostly in newly acquired wild caught specimens. Since they have a direct life cycle leeches can theoretically reproduce in a home aquarium although it is not considered to be likely in most circumstances.

Leeches are easily identified with the naked eye or under low power magnification. Their size, habitat and the presence of an anterior and posterior sucker are diagnostic. In serious infestations, anemia and failure to thrive can occur.

Symptoms

Visible parasites or attachment sites
unexplained sores or wounds
Scratching
Increased respiration and failure to thrive in severe cases

Treatment

Treatment is parasite removal under anesthesia followed by application of topical antiseptic. Organophosphates are also effective for removal. In most instances tank infestation is not considered to be a problem. In cases of infestation, organophosphates will also eradicate the organism but is not recommended for use in the display tank, with invertebrates, or live rock.

Section 4: Gas Bubble Disease

Chapter 1 – Symptoms of "Gas Bubble Disease"

Gas bubble diseases in fish other than seahorses are caused, in the vast majority of cases, by supersaturation of the water with either carbon dioxide or nitrogen (oxygen is easily metabolized, although water *very highly* supersaturated with O_2 will produce the same results).

Although seahorses are just as susceptible to this common form of "gas bubble disease", the causes of the symptoms we generally refer to with the same term are somewhat different, and can occur without supersaturated gases present.

The three types of gas bubble disease are:

External gas bubble disease (EGBD)

This is seen as gas emboli, subcutaneous emphysema – bubbles of gas forming under the skin of the seahorse, usually but not always on the tail. Male seahorses seem more prone to this condition, although it occurs in both sexes.

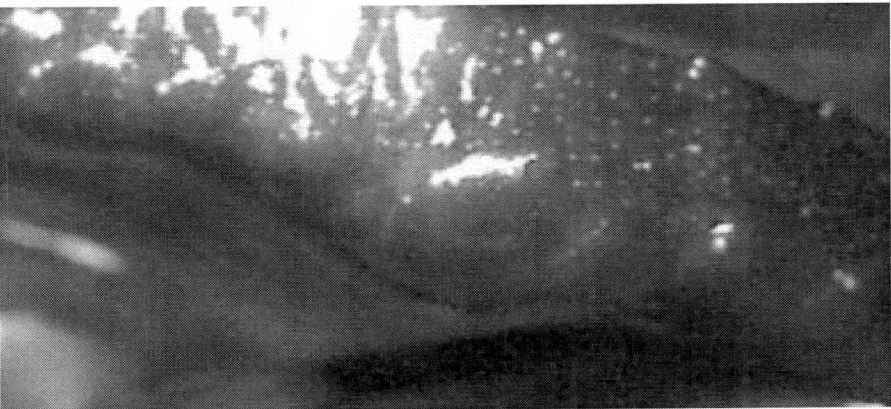

Internal Gas bubble disease (IGBD)

IGBD is the build-up of gases in the abdomen causing positive buoyancy and gross abdominal swelling.

It is worth pointing out that the latter symptom can be confused with fluid retention (edema) resulting from kidney failure as both result in abdominal bloating. However, there is a simple method of differentiating the two conditions. The retained fluid of edema will make the seahorse sink to the bottom of the tank and often lean over under the significantly increased weight. If the animal is floating, it is not fluid retention, and therefore must be IGBD.

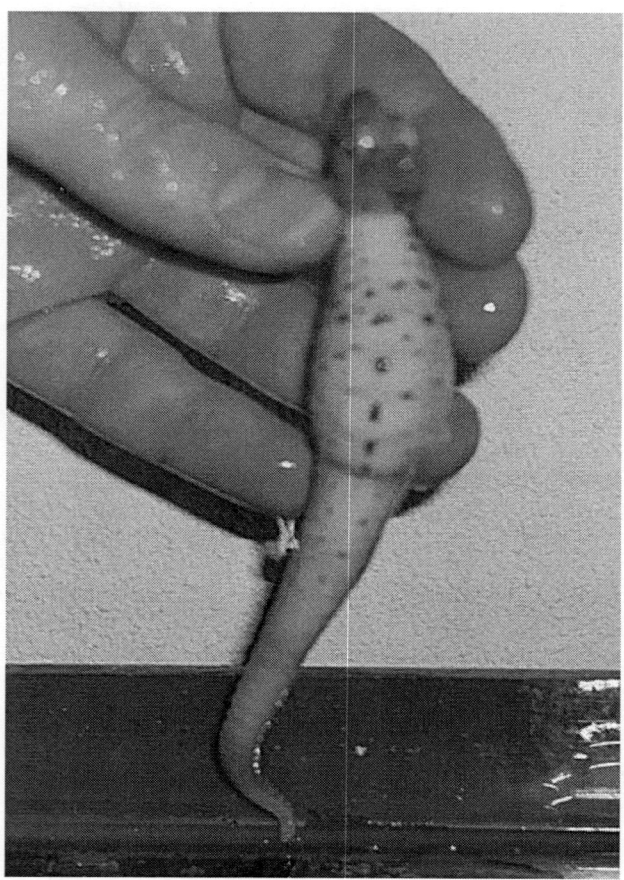

Pouch Emphysema (PE)

This is characterized by trapped gas in the pouch of the male seahorse resulting in a distended, swollen pouch and positive buoyancy (floating). Be mindful that a swollen pouch does not necessarily indicate PE. The seahorse may be pregnant, or just have filled its pouch with water to make it appear larger as part of a courtship display. Only when the seahorse is having problems swimming due to positive buoyancy should he be treated for PE.

Treatments of all forms of GBD

Acetazolamide (trade name "Diamox®") is the treatment of choice for all conditions -EGBD, PE, and IGBD.

Diamox® is a non-bacteriostatic sulphonamide. It will neither kill bacteria as the bacteriocidal potentiated sulfas will, nor inhibit the growth of bacteria as with the bacteriostatic sulfa drugs such as sulfathiazole. Rather, it inhibits production of the zinc-containing enzyme carbonic anhydrase. Its application in treating GBD was pioneered by Dr. Martin Greenwell from Shedd Aquarium in Chicago (pouch emphysema) and Dr. Andy Stamper of The Living Seas (subcutaneous emphysema).

To better understand how acetazolamide treats GBD it is helpful to know the chemical process that is affected by it.

CA (carbonic anhydrase) catalyses the hydration of CO_2 (carbon dioxide gas) with H_2O (water) to H_2CO_3 (carbonic acid) in the following reversible equation:

$$HCO_3^- + H^+ <> H_2CO_3 <> H_2O + CO_2$$
Bicarbonate + Hydrogen <> Carbonic Acid <> Water + carbon dioxide

It is only when the reaction is driven too far to the left or right that problems with the "gas bubble diseases" are thought to occur. When physiological processes cause outgassing in the blood the formation of gas emboli result. This can either be caused by acidosis (lowering of the blood pH), if the reaction is driven too far to the left, resulting in the accumulation of carbonic acid and complications such as a reduction in the blood's affinity for oxygen – or liberation of CO_2 in the most highly vascular parts of the seahorse's body (e.g. the tail or the pouch) where the highest concentrations of carbonic anhydrase occur.

Diamox® is also a diuretic and may help in the cases of edema/fluid retention mentioned previously (abdominal bloating, swelling of the snout, tissue around the eyes and swelling of the tail) and thus has applications for both the conditions that present with abdominal bloating.

Although administered by injection by public aquaria, Diamox® can be administered via a bath treatment or medicated pouch flush by hobbyists, and is generally very effective in treating pouch emphysema and subcutaneous emphysema.

This is the suggested treatment plan.

Bath treatments (EGBD, IGBD)

1) Prepare a hospital tank for the seahorse with a capacity of 10 US gallons. The new water used should be properly mixed, aged and aerated, and carefully pH/temperature/salinity matched to the seahorse's tank so as not to shock the animal.

2) Provide a hitching post and plastic plants for the seahorse to hitch and hide, a heater to maintain the species' required temperature, and provide a couple of airlines to create water movement and aeration. A cycled filter can be utilized, as Diamox® is not toxic to nitrogen fixing bacteria. However, carbon and anything that can remove the medication from the water (skimmers, UV etc) should be removed. If a cycled filter is not available, test the water at intervals for ammonia – even at the lowest detectable levels ammonia is toxic and immunosuppressive. Lighting is not necessary. In humans Diamox® can cause photosensitivity so low levels of ambient light only are recommended.

3) Using the tablet form of acetazolamide (250mg of Diamox®) crush the required amount to a fine powder and mix in with some seawater. There will be a slight residue that will not dissolve. Add the solution to your hospital tank, discarding the residue. This is the dosage required:

Dwarf species (up to 8 cm/3 inches) -¼ 250mg tablet per 10 gallons
Small species (up to 12 cm/5 inches) -½ 250mg tablet per 10 gallons
Medium to large species (over 20 cm/8 inches) -1 250mg tablet per 10 gallons

4) Move the seahorse to the hospital tank and leave in the Diamox® solution for 24 hours. This is day one of the treatment plan.

5) After 24 hours has elapsed, it will be necessary to perform a 100% water change on the hospital tank. This can be done by moving the sick seahorse to a holding container, discarding the water from the hospital tank and adding in the fresh temperature/salinity/pH matched seawater. Prepare the dose of Diamox® and add it to the water. Place the seahorse back into the hospital tank. This is day two of the treatment plan. This process is repeated for up to a total of five days of Diamox® treatment.

NOTE: No side effects or problems have been reported to date from performing 50% daily water changes instead of 100%. 100% changes are recommended because it is currently uncertain how quickly the drug is inactivated by saltwater. This is the safest method, avoiding any potential cumulative dosage. However, 100% water changes are more stressful on the animal, especially if improperly pH/SG/temperature matched, and may not be necessary.

6) If the symptoms abate by day three, treatment can be stopped. However, it is fairly common for the treatment to take longer to have any noticeable effect. Remain patient. Do not rush to lance the bubbles. At the end of the treatment the seahorse may be returned to the main tank. It is currently believed that environmental factors such as poor gas exchange may be a trigger for the disease and it is recommended that this be addressed before returning the animal, or the symptoms may reoccur.

Further notes:

- In humans, loss of appetite sometimes occurs as a side effect of Diamox® treatment.
- This method of treating Gas Bubble Disease has been proven very effective in a series of tests run with sick seahorses under the control of the emergency team from the website www.seahorse.org.
- The repeated doses of Diamox® are required, as the medication appears to lose its effectiveness over 24 hours.
- Toxicity has not been determined, although tests have been run with 500mg Diamox in 30 litres seawater changed daily over a 4 day test period with no apparent ill effects on 2 *Hippocampus reidi* (medium seahorses)
- The long-term effects of acetazolamide on other inhabitants of the marine aquarium are unknown, so the use of a hospital tank is considered mandatory. Additionally, as the condition may have an environmental trigger, removing the animal to an isolation tank is a sensible precaution. In some cases, simply doing this alone can cure the condition without medication.
- Do not mix medications without seeking advice. Do not combine this treatment advice with any other.

Pouch Evacuation

Evacuation of the brood pouch of a male seahorse may be necessary if there is an accumulation of gas in the pouch, and this gas prevents the seahorse from swimming normally. This is actually a simple procedure, and the benefit to the seahorse is immediate.

A blunt tipped bobby pin, soft plastic pipette, or IV catheter sleeve with needle removed will be required.

You need to be able to hold the seahorse firmly underwater. Some people prefer to do this in a separate container (with tank water) but the procedure may be done in the tank. If it is done in a separate container, make sure there is enough water to fully cover the seahorse from back to tip of snout while holding him.

Hold the male so he is lying across the palm of the hand with his upper body resting between the index and middle fingers (as per the photograph opposite). You will usually find that he will wrap his tail around your pinkie finger. Once he has a firm grip on your finger he will feel secure and will usually not let go. If he does not wrap his tail around your finger he may use his tail to grab his head or snout or try block access to his pouch. He may struggle for a bit, but if he is kept immobilized, he'll soon calm down.

It seems that a 45 degree angle is best, as it brings the pouch opening up to the highest point to help expel the air.
Referring to the photo opposite, the pouch opening (indicated with the pencil) is uppermost on the seahorse if held in this position. The tail is wrapped strongly around the pinkie finger, and out of the way. The head and body are held between the fingers, leaving the pouch perfectly positioned for medical attention.

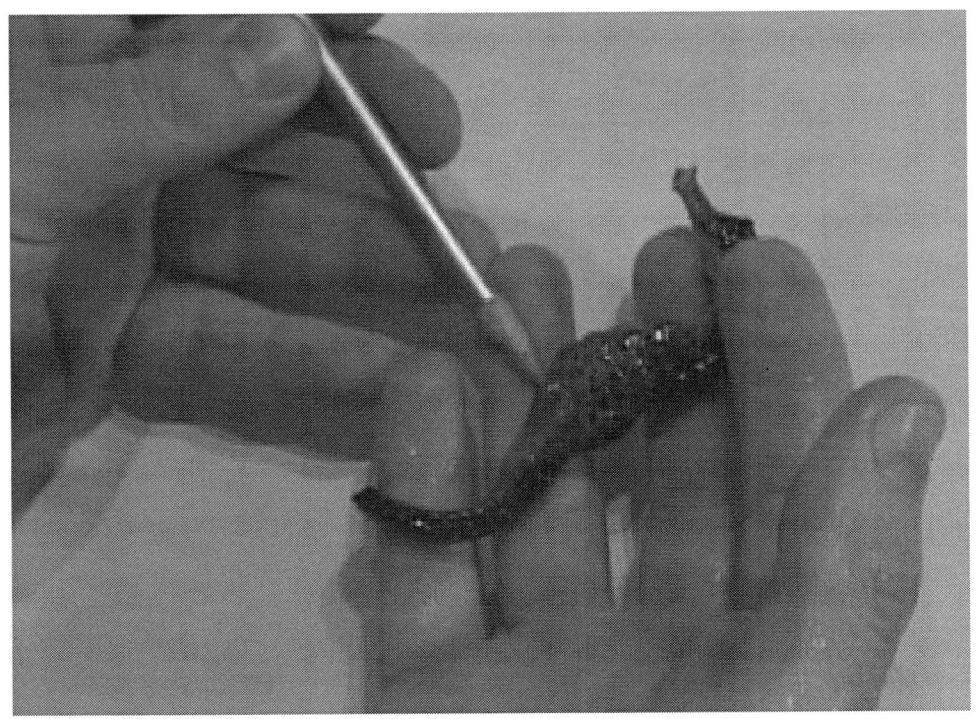

Once he's still, pressure should be applied to the pouch. First try without using any tools to open the pouch. Apply pressure from the front, at the base of the pouch, and work upwards. Do this in sort of a rolling fashion, keeping the area with previously applied pressure downwards and working upwards.

A common mistake is the use of rubbing since the term "pouch massage" is often erroneously used to describe this procedure. Don't do this.

The seahorse pouch is a soft delicate skin covering, so rubbing or massaging the skin can lead to skin damage, infection, and death!

Instead, apply pressure in a direct pushing fashion. It's unnecessary (and dangerous) to move the skin of the seahorse. Apply pressure on the pouch only. It may also be necessary to gently squeeze the sides of the pouch to direct the air towards the opening, using the free hand to squeeze while using the thumb of the "holding hand" to push the pouch as well.

Bubbles should be observed coming out of the pouch opening.

If the air does not come out on its own, it may be necessary to open the pouch.

Hold the insertion tool (blunt bobby pin, pipette, or catheter) between the thumb and index finger of the free hand and work at pushing the sides of the pouch opening away.

The seahorse will be holding his pouch opening closed, so it will take a little while of gently working at it to get him to relax enough to let it open. The opening may not be plainly visible since the pouch folds around itself here. Work gently, careful to not actually insert any object into the pouch while doing this procedure. It's quite clear when it's open as a stream of bubbles comes out of the pouch. Hold open and with the fingers putting pressure on the pouch, keep working upwards until the air is all out.

In a small number of males, even teasing the pouch open will not work to get the air out. In those cases, it will be necessary to use an IV catheter sleeve to insert completely into the pouch to release the air. Slide the catheter into the pouch at an angle so the catheter is parallel to the body/tail. It may require gentle wiggling in all areas inside the pouch to find every bubble of trapped air.

Pouch Flush

In cases of recurring pouch emphysema, Diamox® can be administered as a solution injected into the pouch via a narrow gauge irrigating cannula or plastic 26 or 28 gauge IV catheter sleeve attached to a 0.5 or 1ml syringe.

Using a blender, mix ¼ of a 250mg tablet with a cup of seawater. Fill the syringe with about .5ml of this solution, avoiding the residue at the bottom of the cup. The seahorse should be held as per the procedure for pouch evacuations.
Insert the catheter sleeve slowly and gently a small way into the pouch opening and inject this solution slowly into the seahorse's pouch, leaving the solution in the pouch. Make sure you are familiar with the location of the pouch opening.

Never use a metal needle for this procedure.

The procedure may have to be repeated twice to be effective. In stubborn cases, it is recommended to concurrently administer broad-spectrum antibiotics. Diamox® and antibiotics have been used simultaneously and successfully without apparent side effects.

This can be an effective treatment but it can be tricky, especially with smaller seahorses. It should be noted that, as with any such procedure, the treatment carries a risk of physical injury if done incorrectly.

Section 5: Treatments

Chapter 1 – Dips, baths and prolonged immersion

Depending on the duration of the treatment, waterborne treatments are referred to either as baths, dips or prolonged immersion.

Dips are generally very short term – 5 to 15 minutes - as prolonging the treatment would harm the animal. An example of this is a freshwater dip.

Baths are slightly longer duration – about 30 minutes to a couple of hours (e.g. a 45 minute formalin bath).

Prolonged immersion treatments range from 24 hours to ten or more days. These are generally waterborne antibiotics, which require longer-term administration to be effective.

Formalin bath

The product used is Formalin 3 (37% formaldehyde).

What you will need:

A clean, never used before ten liter (about 2.5 gallon) bucket or
similar food safe container or tank (transparent containers afford
greater visibility for observation of the animal during treatment).
Formalin 3 (37% formaldehyde)
Fresh saltwater, aged and aerated, matched to the same
salinity/specific gravity, temperature and PH as the tank the seahorse
is currently in to avoid shock.
A pump and an airline

Instructions:

1. Heat the seawater to the same temperature as the tank the seahorse
is in, and PH match using a PH buffer if necessary. Fill the container
you are going to use for the bath with the seawater.

2. Make sure the formalin does not contain white sediment. This is
paraformaldehyde and is highly toxic. **If the bottle contains
sediment, do not use it.**

3. Formalin 3 is dosed at 200ppm for 45 minute to hour-long baths.
Using a measured pipette or syringe, add formalin 3 at the rate of
1ml per 5 litres. Be careful not to breathe the fumes.

4. Add the airline and something for the seahorse to hitch to. You
must use an airline with the bath - formalin removes oxygen from the
water (each 5 mg/L chemically removes 1 mg/L dissolved oxygen).

5. Put the seahorse into the container and observe continually during the treatment for signs of distress such as thrashing, twitching etc. Most seahorses tolerate formalin treatment well but if you see any significant signs of distress, stop the treatment. Maximum duration is one hour.

Sometimes you may see parasites leaving the host in cases of gill infestations. If you have a microscope, collect the sediment from the dip using a fine micron mesh and examine.

Recently there have been questions raised as to whether formalin treatments are dangerous even at recommended dosages. However they are still commonly used by public aquaria in this manner to treat for ectoparasites.

Formalin is a gill irritant and is more toxic at higher temperatures. It is not an effective treatment for fish with ulcers, and should not be used with fish that have been very recently shipped/transported.

Methylene blue

Methylene blue, or "meth blue", is a dye that forms a deep blue solution when dissolved in water.

It is a mild antibacterial and antiparasitic (this action is thought to be due to binding with cytoplasmic structures within the cells of the pathogens). It is effective in reversing nitrite toxicity. Environmental nitrite causes hemoglobin in the blood to be oxidized to methemoglobin, which cannot deliver oxygen to the tissues (a condition known as Methemoglobinemia which leads to the body being starved of oxygen) even when sufficient dissolved oxygen is available environmentally.

Methylene blue reduces methemoglobin to hemoglobin, and increases the blood's affinity for oxygen. If treated promptly, and the condition is not too advanced, fish should make a full recovery quickly. It also has applications in treating cyanide poisoning and effects of other environmental contaminants. Proprietary stock solutions ready for use are readily available from local fish stores and online retailers. To treat, simply follow the instructions on the packaging. Methylene blue has a wide safety margin and is non-toxic when used as recommended. Fish tolerate relatively high dosages without side effects and dosing is not overly critical within sensible limits. Treatments usually involve short 1-4 hour baths or up to 2 days in the case of nitrite toxicity. Visibility is reduced - this is normal and will not harm the fish.

Under no circumstances should Methylene blue be used in the display tank. As well as being a potent dye that will stain anything it contacts, it is toxic to biofiltration bacteria.

Malachite Green

Proprietary preparations of Malachite Green are readily available from local fish stores or online.

Care should be taken to find a product indicated for use with saltwater fish and follow the recommended treatment instructions from the manufacturer, as it is commonly sold as a preparation for pond use, which may be of a higher dosage.

Whether this is an appropriate medication for seahorses is debatable. Though it has been used with some success to treat syngnathids by public aquaria , there are reports of toxicity with scaleless fish and it is a respiratory poison. As such it should only be used with caution, to treat parasites such as Uronema, Brooklynella etc.

Freshwater dips

Freshwater dips should be performed in a separate, clean container or hospital tank with previously aerated reverse osmosis or dechlorinated water buffered to the same PH as the water in the main tank, and heated to the same temperature, but without any salt mix added.

It is best to use a glass container so you can see the horse's reaction clearly although a clean, never used before 10 liter (about 2.5 gallon) bucket will do. Hitches for the seahorse should also be provided.

Observe the seahorse for the duration of the dip, which should be ten minutes.

Some immediate signs of distress or shock may be seen. Sometimes the horse will immediately lie on its side on the bottom. If this happens, tap it gently with your finger several times. If it does not respond, stop the treatment. Most horses tolerate the treatment well. If you see continued signs of distress – such as twitching, thrashing around, stop the treatment. Never fresh water dip a seahorse with an open wound, as the body fluids can be drawn through the wound via osmosis.

Sometimes you may see parasites leaving the host in cases of gill infestations. If you have a microscope, collect the sediment from the dip using a fine micron mesh and examine.

Hyposalinity

Hyposalinity treatment refers to lowering the salinity of the water for a prolonged period of time – usually enough to cover the life cycle of certain ectoparasites, which may be as long as four to six weeks.

Ordinarily this should not be done in the main tank as invertebrates such as snails and shrimp cannot tolerate lowered salinity.
The treatment is harmless to the seahorse and may be combined with antibiotics in case of secondary bacterial infection from ectoparasite infestation. A study into the effects of salinity on *H. kuda* juveniles concluded "Juvenile *H. kuda* are able to survive in dilute seawater (15 ppt) for at least 18 days without any compromise in growth (both wet and dry body weight), survival, and total body water".

Prepare the hospital tank as mentioned previously, with a cycled bio filter. To be effective, the treatment must be 14-15ppt (parts per thousand) or 1.010-1.011 SG (specific gravity). It is important to measure this with a calibrated refractometer. These can be purchased cheaply online. They are much more accurate than swing arm hydrometers, which may give a very different reading from the actual salinity – and may cause the treatment to fail.

The seahorse does not need to be acclimated to the lower salinity, but the water must again be temperature and PH matched, and properly mixed, aged and aerated, and if tap water is used for source water, it should be treated with a product that neutralizes both chlorine and chloramines.

Hypersaline dips

The reverse of hyposalinity treatments, hypersaline dips involve placing the seahorse in seawater with increased salinity for a short space of time.

The treatment should be performed in a separate, clean container or hospital tank with previously aerated reverse osmosis or dechlorinated water buffered to the same PH as the water in the main tank, and heated to the same temperature, salt mix added to 1.040 (specific gravity) or 53 ppt (salinity), then mixed with a powerhead and aged (do not use immediately as undissolved salt mix will damage the seahorses' gills)

It is best to use a glass container so you can see the horse's reaction clearly although a clean, never used before 10 liter (about 2.5 gallon) bucket will do. Hitches for the seahorse should also be provided.

Observe the seahorse for the duration of the dip, which should be ten minutes.

Most horses tolerate the treatment well. If you see continued signs of distress – such as twitching, thrashing around, or lying on the bottom, stop the treatment.

Chapter 2 - Treatment of ectoparasites

None of these treatments, with the possible exception of hyposalinity in very basic, mainly artificial setups, should be performed in the display tank.

Osmotic Shock Therapy

OST targets ectoparasites, such as protozoa, that cannot osmoregulate. It can be effective on external infections and gill, snout, and trigger infections but is not appropriate for treatment of endoparasites.

This includes the following:

1. Freshwater dips
2. Hyposalinity
3. Hypersaline dips

Biocides

These also target external parasites. Some of these are very harsh chemicals, prolonged exposure to which will injure the seahorse.

1. Formalin
2. Malachite Green

For larger parasites such as metazoans and parasitic copepods, organophosphates such as Trichlorphon (Dimethyl-2,2,2-trichloro-1-hydroxyethylphosphonate) can be employed. Once again, these are also harsh on the animal being treated.

Chapter 3 – Specific treatments

Hunger strike

The number one most effective treatment for a hunger striking seahorse is to feed live foods. Do NOT tube feed a seahorse without first attempting to feed it with appropriately sized live foods such as ghost shrimp or mysids, amphipods, or possibly guppy/molly fry, or even vitamin/carotenoid/beta glucan enriched adult artemia (though the last two may not be recognized as food, especially by wild caught seahorses who will never have encountered them before). By far the best choice of food to stimulate a hunger-striking seahorse is the white shrimp *Penaeus vannamei*. This shrimp is available to hobbyists in the US from www.seawaterexpress.com. They are available in various sizes and are guaranteed pathogen free.

Sometimes the seahorse will not be able to actively hunt the relatively fast moving shrimp. One trick that can be employed is injuring ghost shrimp so that they are still alive and moving, but unable to escape. While this sounds gruesome, it is no worse than the fate awaiting it once the seahorse gets hold of it. Using plastic feeding tongs, squash the shrimp along the tail just after the carapace. Leave it alive and moving (and thus triggering the feeding response) but incapacitated and easy prey for the seahorse. This is often very effective for hungry seahorses that are sluggish for a variety of reasons including recovery from a recent illness.

Force Feeding (tube feeding)

This is NOT a quick cure for wild-caught seahorses that will not eat frozen foods. This is only for serious use, when an injury or illness threatens the life of the seahorse. If the seahorse is refusing to eat for four days or more, the following procedure should be considered necessary.

Prolonged use of this procedure will kill a seahorse. But it will quite often save a sick one.

SUPPLIES:

0.78mm catheter

syringe x 2

cod liver oil (omega 3 oil)

spirulina flakes/powder

frozen mysis shrimp

100 micro sieve (actually a 250 micron will do)

clove oil

container of seawater

mortar and pestle (a grinding tool)

1) Insert the needle into the oil capsule and draw .2 to .3 cc's of oil

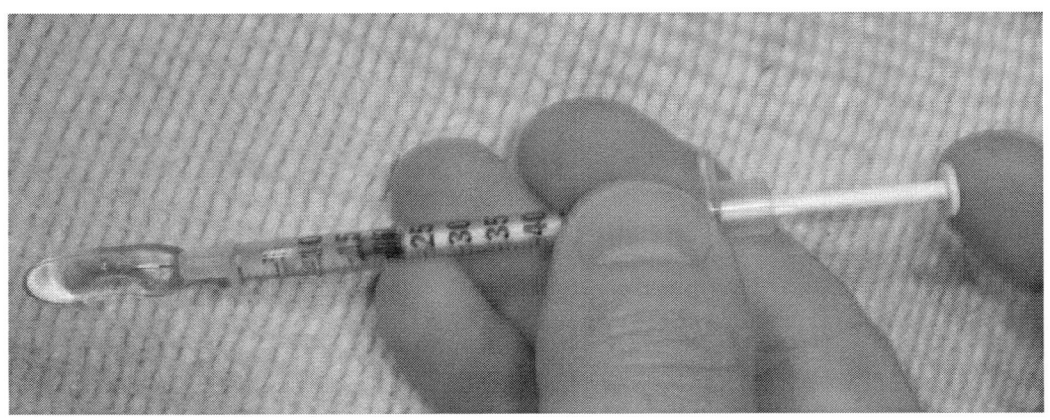

2)   Add the Omega 3 oil and spirilena powder into a cube of
     mysis shrimp. Don't use brine - it has very little nutritional
     value.

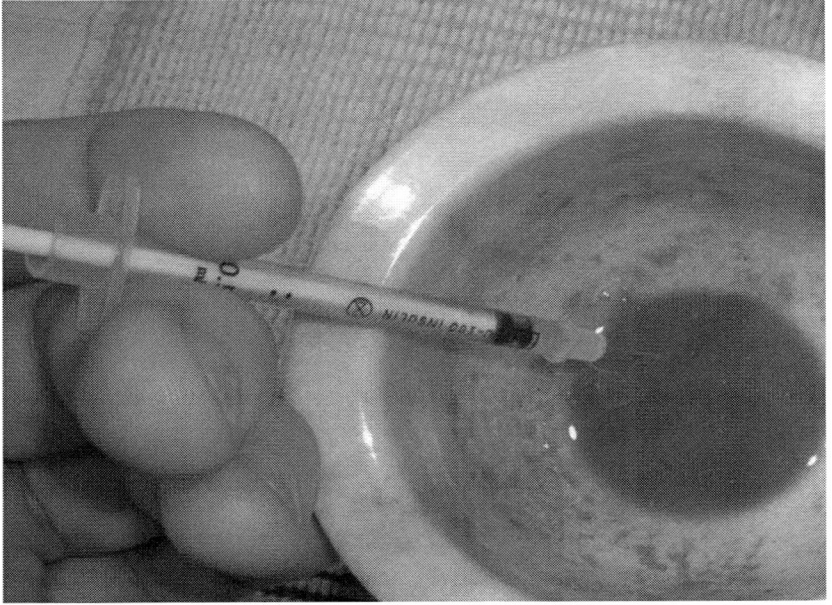

3)   Grind the contents into a lump-free, fine consistency.

4) Pour the solution through a 100 micron sieve to remove lumpy bits that did not grind out properly.

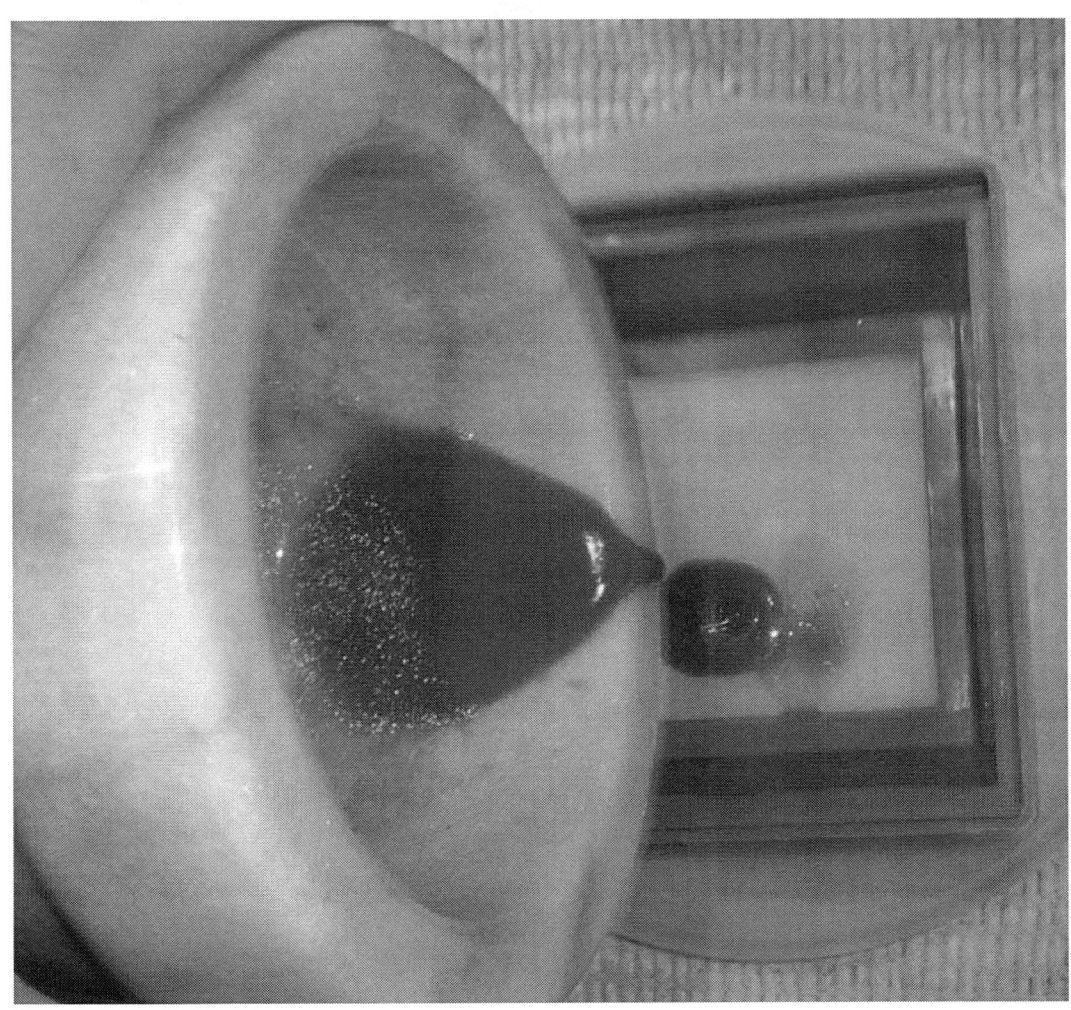

5) Using the needle, draw 1 cc of the solution into the second syringe (using the needle to draw the solution prevents blockage when feeding back through the catheter).

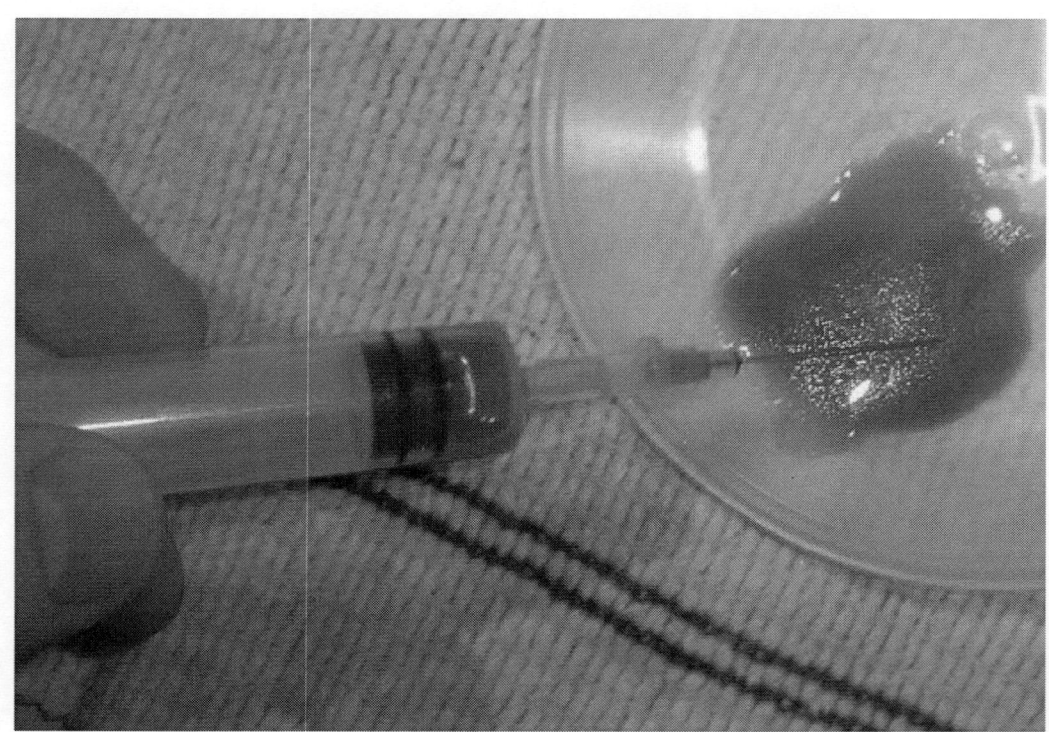

6) Place the catheter onto the syringe

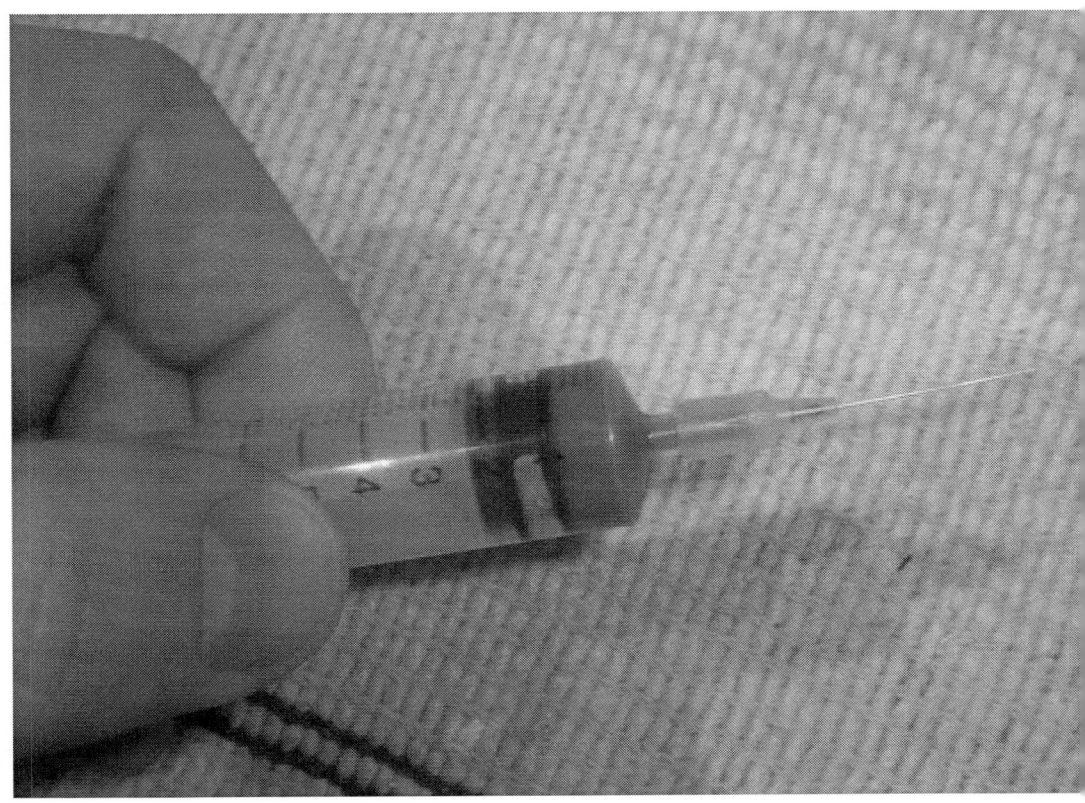

7) Add one drop of clove oil into a bowl of seawater. Use tank water and throw it away afterwards. Stir the water and oil so that it mixes (yes, it will).

8) Place the seahorse into the bowl. The clove oil in the water quickly subdues the seahorse and he'll get sleeeeeeee…….. This step MUST BE DONE QUICKLY. The seahorse is actually dieing under the clove oil anesthetic.

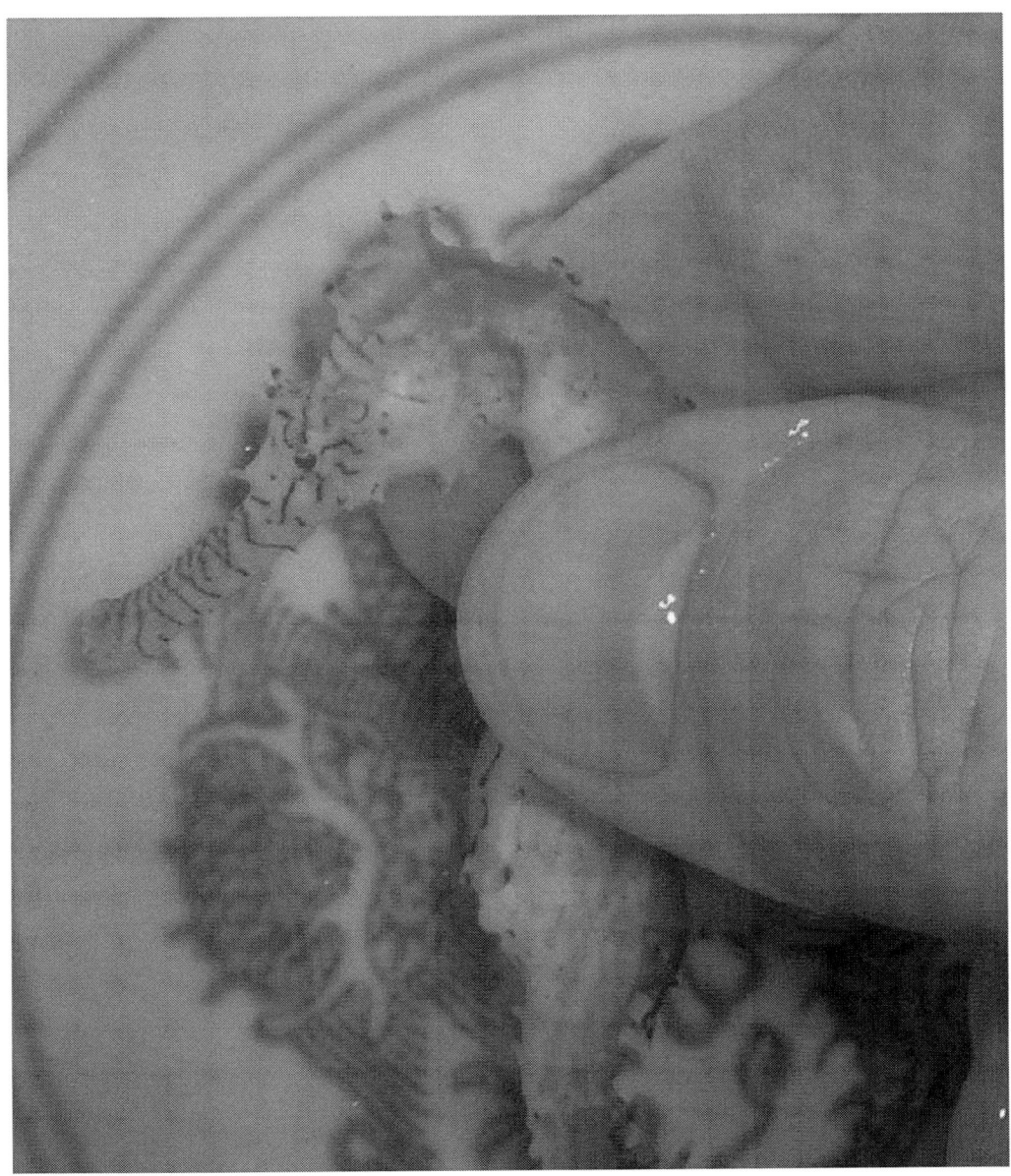

9.) Insert the catheter into the snout. Move the catheter and the seahorse head to get the catheter tip past the gills and to the back of the palate. Then SLOWLY inject the 1 ml solution into the seahorse. Most of this will actually go into the seahorse's throat. This is why the seahorse should be sedated.

Remove the catheter. Now the seahorse can be revived. Place the fish in a bowl of fresh half strength seawater that is heavily oxygenated. The fresh water removes the clove oil and the seahorse revives. If this does not happen almost immediately, then use the syringe with NO NEEDLE and GENTLY squirt water down the snout, which then goes out the gills and the seahorse revives. Place the seahorse back into the hospital tank.

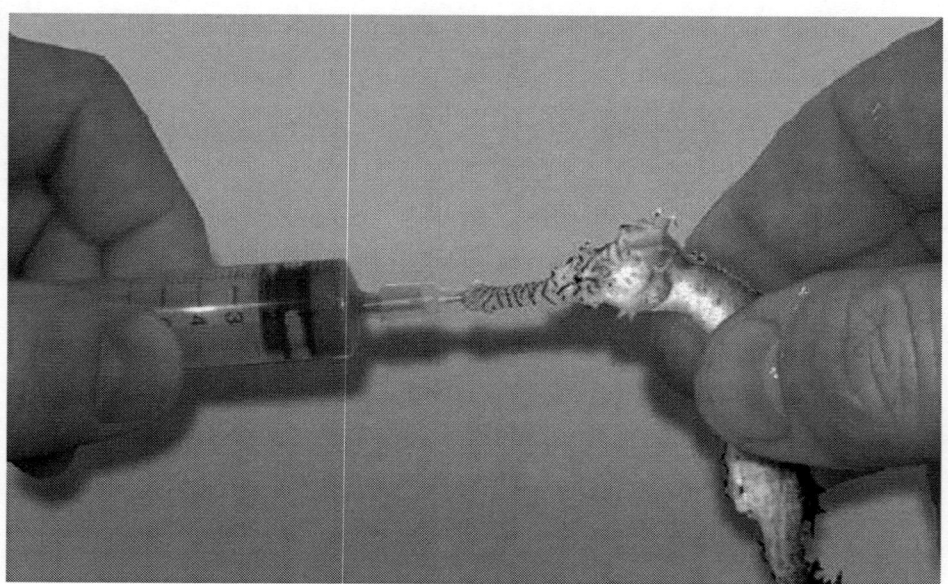

Chapter 4 – Physical injuries

Tail Lesions

The seahorse tail is a unique modification of the caudal fin. It is prehensile and mostly used for maintaining position. During pursuit of prey, it enables the seahorse a stable platform from which to launch its lightning fast attack, also known as "snicking". The tail is constantly dragged over the substrate and coral which leaves it open to constant trauma and/or abrasion. In addition, the tail is furthest from the heart which correlates with low oxygen tension and the proclivity for lesions at this site.

The seahorse tail is an essential part of the animal's daily life. It is used in getting food, mating, and a number of other activities. Damage or disease in this area can lead to the inability or lack of desire to eat, leaving the well being of the whole organism in jeopardy.

Tail lesions can appear as ulcerations, discolorations, or loss of color and as nodules. The ulcerative lesions are considered to be mostly due to bacterial infections but could also be caused by protozoans such as uronema. Vibrio is the most common bacterial pathogen in seahorses so treatment is usually an antibiotic with coverage for this organism. If in doubt, a wet prep smear can be used to try and rule out an organism like uronema but identification is often difficult.

Nodular lesions are very common. Clear, gas filled vesicles are usually diagnostic of external gas bubble disease. They can be accompanied by internal gas bubble disease or can occur separately. Treatment with Diamox usually alleviates symptoms.

Nodular lesions can be more solid or cystic in nature. Cystic lesions filled with pus are called abscesses and are usually due to Vibrio infections although other bacterial infections are possible. Lesions that tend to be more solid in nature include Glugea, Nocardia, and Mycobacteria. Since there is no effective or good treatment for these maladies, quick isolation is essential. Glugea requires biopsy and identification under strong magnification for a reliable diagnosis. The nodules are called xenomas and are somewhat cystic, filled with microspores which are released when the lesion ruptures at maturity. Nocardia and Mycobacteria can be cultured in the lab but are slow growing and most easily identified by gram stain or acid fast stain.

Loss of color in the tip of the tail also occurs. It is a slowly progressive disease that if left unchecked can involve the entire tail and eventually the rest of the animal. At this time it is thought to be caused by a bacterial infection which interferes with the blood supply to the area, causes localized death to the tissues, and the pale white color. Treatment for Vibrio is recommended for these lesions.

Chapter 5 – Other treatments

Garlic

Preparations of garlic extract designed for use in supplementing aquarium foods, such as Kent's Garlic Xtreme, are anecdotally reported to stimulate feeding response in seahorses. Unfortunately this does not always work, and when it does it may be co-incidental.

Beta glucan

Disease does not just occur because pathogenic bacteria, parasites, or viruses are present.

There are numerous complex interactions between the seahorse, the pathogen in question, and the environment. The conditions have to exist for the infection to take hold, and that involves a combination of genetics, the general health of the seahorse, nutrition, and environmental stressors.

Fish have natural defenses against pathogens in the form of the specific and non-specific immune system.

Beta glucan (B-1,3 D polyglucose, a molecule derived from baker's yeast) stimulates the production of macrophages, immune cells that are found all over the body of the seahorse, that consume foreign substances such as bacteria, as well as dead cells and other debris.

In this manner, Beta glucan can help the seahorse fight disease by stimulating the immune system. While there is debate over continuity of results, yeast glucans are presently the most commonly used immunomodulators in aquaculture, with applications ranging from quarantine and prophylaxis to use as an adjunct to antibiotics and vaccination.

Professional aquarists use a soluble Beta glucan. These are proprietary treatments that are administered as a short-term bath or prolonged immersion treatment, with the dosage for the former being around four times that of the latter. As with all such treatments, the instructions for administration and dosage vary and (with the exception of certain over-the-counter fish antibiotics) the procedure outlined by the manufacturer should be followed.

Liquid soluble forms of Beta glucan are highly preferable, but often prohibitively expensive and difficult to obtain. An alternative is capsule and tablet forms that are available from health food stores and online retailers. Capsules contain powdered Beta glucan. The tablets must be ground up into a fine powder with a pestle and mortar or similar.

This non-soluble form must be administered orally via bio-encapsulation. In plain English, that means getting something that the seahorse is going to eat to eat the Beta glucan first. This can be achieved by either using live adult artemia or live ghost shrimp or similar.

Adult artemia should be vigorously aerated in a bottle or container with a rigid airline or airstone. The aeration must be enough to keep the powdered Beta glucan in suspension – Artemia sp are filter feeders, and indiscriminately eat anything of a certain size in the water column. Note that the early instars are too small to eat the particles.

Live ghost shrimp can be injected with the powder by mixing it with a little water and using a needle (24-26 gauge) and syringe to inject the solution into the fleshy part under the armoured carapace or abdomen.

It is worth noting that the shrimp may also eat the tablets if they are placed in their tank, which simplifies matters. The powdered form can also be added to a gruel used to tube feed the seahorse.

Beta glucan takes around 72 hours to become fully active when administered orally.

Section 6: Ongoing research

Chapter 1 – Fatty Liver Disease

Fatty liver disease affects many species including humans. In a Seahorse autopsy series of 30 cases, 46.6% were found to have significant fatty liver disease. Cases were obtained in chilled Styrofoam boxes from hobbyists via next day air. In all cases the animals were received within 48 hours of death and had shown signs of sickness leading to death. Most cases were also accompanied by some form of treatment protocol, usually with antibiotics in a quarantine tank.

Upon receipt of the specimens all were photographed, weighed, and measured. They were then bathed for 5 minutes in denatured alcohol in an attempt to sterilize the skin surface for bacterial cultures. Cultures were obtained from the peritoneal cavity using sterile procedure and streaked on sheep blood agar. The gills and viscera were submitted for histological examination and special stains if necessary.

Histologic sections of seahorse liver tissue were evaluated for fat content according to the following scale

0 - No evidence of fatty liver
1- Greater than 5% but less than 33% of the liver showing fatty change
2- Greater then 33% but less than 66% of the liver showing fatty change
3- Greater then 66% the liver showing fatty change

Cases considered to be too autolyzed for evaluation were excluded from the study. Since the distribution of disease in the affected livers was found to be patchy and zonal, grading was conducted by analyzing the ratio of the number and size of affected zones to the number and size of unaffected zones. No consideration was paid to the age, size weight or species of seahorse examined. Significant disease was judged to be grade 2 or greater.

After analysis the following results were recorded.

Grade	# of animals	(%)
0	15	(50 %)
1	1	(3.3 %)
2	8	(26.6%)
3	6	(20 %)

Fatty liver disease in humans has been related to high fat diets, drugs, alcohol, diabetes, obesity, vitamin deficiency, excess vitamin A, and starvation, just to name a few. In domestic and captive animals the cause seems even less clear. It has been identified in domesticated parrots, llamas, alpacas and fish. In llamas and alpacas it is believed to be related to stress and/or diet. In fish the known causes include.

1. Biotin and/or choline deficiency
2. Diets low in unsaturated fats (EPA, HUFA, DHA)
3. Vitamin E deficiency
4. Auto-oxidation of feed
5. Excess or the wrong kind of dietary fat
6. Damage from toxins

Biotin and choline deficiency have long been known to cause fatty liver disease in fish. Diets low in unsaturated fats (essential fatty acids) and fatty diets are self evident, well researched, and logical.

Auto-oxidation of feed is seen to occur mostly in farmed salmon and trout. The condition develops when high-energy feed loaded with polyunsaturated fat becomes oxidized due to any cause such as improper storage. The unsaturated fats can oxidize (rancidification) forming peroxides, aldehydes and similar toxic compounds. Exposure to even small amounts of these compounds can produce kidney damage, liver enlargement, and anemia. Supplementation of such rancid feed with Vitamin E can prevent disease from developing. This helps because during the oxidation process dietary vitamin E is destroyed.

In the seahorse, the liver is divided into two major lobes with one being situated in the abdominal cavity and the other being located either in the abdomen or in the neck region. Generally, depending on the size of the specimen, the gall bladder measures a few millimeters in diameter, is prominent, and is filled with dark green bile. Connections to the bile and pancreatic ducts, deliver their contents into the gastrointestinal tract.

Microscopically, the liver in seahorses is a combination of liver cells and pancreas and can be considered to be a hepatopancreas. The hepatic sinusoids and biliary canaliculi are fewer than in mammalian species. Portal tracts are not present. Bile canaliculi originate on an intracellular level to eventually fuse into a proper bile duct that fills the gallbladder.

Dispersed throughout the liver tissue are islands of pancreatic tissue, some showing secretory glandular formations containing numerous goblet cells filled with zymogen granules. As an organ, the seahorse liver is second in size only to the seahorse's skin and is the largest internal organ.

Some known functions of the seahorse liver include:
Detoxification
Cholesterol metabolism
Bile Production
Fat and Carbohydrate Metabolism
Protein Metabolism
Energy storage

The job of a seahorse liver is difficult and varied. It must detoxify harmful substances taken in through feed and through the gills. Cholesterol metabolism and formation is regulated by the liver. Cholesterol is the basic component of both hormones and bile. Therefore the liver plays a role in hormone formation and regulation. Bile is necessary to absorb dietary fats and vitamins from the diet. Carbohydrates and fats absorbed in the intestines are processed, stored and repackaged in the liver for energy and various other uses all over the body. Proteins such as albumin are made in the liver to provide the building blocks for growth and repair. The liver can also transform unused proteins into glycogen or fat for storage. Anticoagulants or clotting factors needed to control hemorrhages are also made in the liver. It appears that the liver of a seahorse has a big job to do and this is coupled with the extra burden of also functioning as a pancreas.

Theoretical causes of fatty liver in this series include all of those listed above. Many of the animals had been very ill, requiring multi-drug therapy over long periods of time. This chronic disease state in itself could cause fatty liver despite the effects of the drugs or the poor nutritional state of some of the animals because they were not eating. In none of the animals was there any sign of poor development suggesting there had been a chronic nutritional deficiency besides that produced by anorexia related to disease.

All of the animals appeared to have been well fed prior to or during the time of the illness. Exact data concerning supplementation or enrichment of food items was not obtained.

Since there is no way to rule out either drug effects or disease related anorexia, attention must turn to nutritional excess or deficiency as a potential cause for fatty liver in these seahorses.

Classically it has been taught that seahorses need diets high in HUFAs to survive in captivity. Most of this data comes from studies done using the growth of fry as a benchmark. Fry start off at a few millimeters in size and, in the course of a few months, grow to many centimeters in length. Metabolic demands are enormous and the young seahorses will not thrive unless they are supplied with a steady diet high in fat. As they mature, their steady state fat requirements are reduced but still can remain relatively high if they are actively breeding. Anecdotal evidence with captive lion and scorpion fish fed exclusive diets of freshwater fish such as goldfish has led some to believe that the low HUFFA content and high saturated fats found in freshwater species may also be a cause of fatty liver in saltwater predators.

Studies of plankton populations have shown that the content of HUFAs can vary with the seasons and from region to region as the seasons change. In one study by Lokman, variation as high as 6% was recorded. Also to be taken into account is that as the seahorse grows its prey becomes larger. Larger prey items generally contain lower concentrations of fat than smaller ones. These variations, both seasonal and dietary, provide the adult seahorse with a constantly changing fat content in the diet. In addition, seahorses in the wild probably don't get their fill, are opportunistic feeders, and might not live for more than a year, or possibly two, so they are exposed to less fat over their lifetime than most captive fish.

Seahorse fry on the other hand, wild caught or tank raised, need the constant supply of unsaturated fats to fuel their incredible growth. This has been proven in captivity but it is not known what the rate of fry mortality in the wild is secondary to poor nutrition.

What percentage of fry does die in the wild from nutritional deficiency? Has the goal of good husbandry created a situation where adult seahorses are being fed as fry? Do wild adult seahorses get fatty liver and what is the total fat content of their diet? How often and how much do they eat? These questions cannot be answered until further investigation is undertaken.

For the time being, the best course of action pertaining to adult seahorses is a conservative one. Avoid using fatty acid supplementation more than once or twice a week on adult seahorses. Be sure to supplement food with vitamins once a week. Don't feed seahorses until they look like they are going to burst. Discard old food, frozen or not, because it will be vitamin depleted and possibly oxidized. The same pertains to vitamin and fatty acid supplements. When it comes to supplementation, more is not always better. And seahorses should have one day of fasting per week.

When actively breeding a pair of seahorses, increase the fats in their diets while reducing fats for those who are not breeding. Use live food if available on a routine basis or as a periodic supplement. Do not exclusively feed freshwater fish fry as a staple diet.

Hopefully, continued research in this area will reduce the incidence of fatty liver disease in these delightful creatures.

Chapter 2 – Vibrio

Vibrio Research

The presence of Vibrio in seahorses that are healthy or have died of another cause suggests a carrier state. There have been many accounts of intra and inter species mixing which result in death of one or more previously healthy individuals. Vibrio may play a role in this process and needs further investigation. To begin the process of investigation the following study was performed.

At autopsy, Vibrio is identified quite often in material taken from the peritoneum of captive seahorses. In an autopsy series performed on a mixed population of dead seahorses received from hobbyists, 38 wild caught and captive bred Seahorses were opened under sterile technique and peritoneal samples were submitted for culture on sheep blood agar at 25 C. Before the cultures were taken, the specimens were soaked for five minutes in denatured alcohol. Incisions were made with sterile instruments at the keel. Effort was made to only touch abdominal organs with the culture swab, avoiding the sides of the incision.

Of those samples taken, only 11 (29%) showed no growth. All distinct bacterial colonies growing on the plates were worked up, isolated and identified using salt supplementation in a Microscan Walkaway 40 at 35C. Exact identification of Vibrio species did not always occur with some being reported as Vibrio NOS (Not Otherwise Specified) due to poor growth. All isolates were identified with greater than 90% confirmation. For the samples with growth the following isolates were found:

H. erectus (15 total cultured, 9 positive)
Vibrio NOS
V. alginolyticus
V. fluvialis
V. parahaemolyticus
V. vulnificus
V. hollisae
V. damsela

H. reidi (17 total cultured, 12 positive)
Vibrio NOS
V. alginolyticus
V. cholera
V. parahaemolyticus
V. damsela

H. barbori (2 total cultured, 2 positive)
V. alginolyticus

H. comes (1 case, positive)
Vibrio NOS

H. kuda (1 case, positive)
Vibrio NOS

H. abdominalis (2 cases cultured, 2 positive)
Vibrio NOS
V. hollisae

The clinical presentation and histologic material was reviewed for all cases and the cause of death was determined. Attention was paid to signs and symptoms during the illness such as body ulcerations or erosive skin lesions. Histologically, bacterial abscess formation in a major abdominal organ was considered diagnostic of Vibriosis, with the liver and kidney being the two most affected organs examined, either alone or in combination with other sites.

General Autopsy findings

Some of the seahorses that had not died of Vibriosis, or were not cultured and used in this study, died from identifiable causes. Worms were a problem mostly for wild caught *H. erectus*. They did also appear in all non-erectus wild caught species but were usually small in size and number, being located in the wall of the gastro-intestinal tract or at the base of the gills. In only one case were worms identified to actually be in the lumen of the gastro-intestinal tract. Worms were not identified in captive bred species.

In six cases wild caught erectus were identified with major infestations leading to death. These worms were of a larger size and appeared to roam freely in the abdominal cavity laying elliptical eggs which contained small larvae. The gas bladder was a favorite site for this activity which caused clinical signs of buoyancy instability in some of the horses. In one chronic case, a mass formed to replace the air bladder entirely. Sometimes the large abdominal worms were also seen with a smaller parasite in the gills. Whether the two are related is not known. This animal grew out three strains of Vibrio which were considered not to be the cause of death.

One reidi died of a combination of parasite load and old age. The remainder were either unexplained deaths or mass deaths with gill damage or appropriate history, the result of what could be called bad water (tanks crashing, toxins, bad source water, etc). The animals not cultured for the study were considered too decomposed, usually due to being delayed in transit, but were still submitted for histologic tissue analysis.

Kidney lesions were identified in both erectus and reidi. Findings include: stones with dilated urine collecting system, evidence (hemosiderin deposition) of partially resolved inflammatory disease, infestation by worms (erectus), and Vibrio abscess formation.

Analysis of Vibrio Data

H. erectus cultures were positive for Vibrio 60% of the time. One animal cultured out 2 Vibrio strains (NOS, fluvialis) and one cultured out three strains (vulnificus, hollisae, fluvialis). Of the 9 erectus with positive cultures, only three were determined to have died from it. In one case with positive cultures, the cause of death could not be determined and in the remaining 5 cases the cause of death was parasites. Of all erectus cultured only 20% died from Vibrio and in those with positive cultures 33% appeared to have died from the disease. The cause of death in those with negative cultures was either parasites or water quality issues. The 3 erectus that died from Vibrio cultured V. alginolyticus in one case, V. NOS and V. fluvialis in the second and V. NOS in the third.

H. reidi cultures were positive 70% of the time. Three animals cultured out more than one isolate with 2 of them producing 2 isolates (alginolyticus, damsela and cholera, parahaemolyticus) and one with three isolates (NOS, alginolyticus, parahaemolyticus). Of the 12 with positive cultures 6 were determined to have died from it. In the remaining six the cause of death was undetermined. Of all reidi cultured 35% appeared to have died from Vibrio. Of the six reidi that died from Vibrio, 2 cultured V. NOS and the other 4 died from *V. alginolyticus, cholerae, parahaemolyticus* and *damsela,* respectively.

In the 2 *H. barbouri* the cause of death could not be determined. The *H. kuda, comes* and *H. abdominalis* all died of the disease. From these results it appears that Vibrio is a common, sometimes pathogenic, organism found in captive seahorses.

The data is especially useful in the case of the reidi and erectus. While *H. erectus* quite commonly harbor the bacteria (60%) only 33% positive for Vibrio species actually die from it. Of all reidi and erectus in the study, *H. reidi* were more likely than erectus to show positive cultures (70%) and also more likely to have died of the disease (Erectus 20%, reidi 35%).

It appears that a carrier state does exist for both *H. reidi* and *H. erectus* while that is less clear for other species due to the low number of animals received. At autopsy, *H. erectus* was also found to be more likely to harbor significant parasites than all other species examined. Perhaps different strains of vibrio are the causative agent in mortalities seen when mixing different species or seahorses from different regions.

Additional studies are necessary to determine if seahorses in the wild culture positive for Vibrio or is it just a manifestation of captive animals. Mixing studies done with tagged Vibrio strains could demonstrate disease transmission from one seahorse to another and further define the importance of a carrier state.

Chapter 3 – Soft Plate Disease

Soft Plate Disease

Reproduced with permission from Pete Giwojna.

Seahorses and pipefish that receive a diet deficient in calcium may be prone to "soft plate" syndrome, which is a progressive disease characterized by decalcification of the bony plates that fuse together to form the exoskeleton (Greco, 2004). In the olden days, seahorses fed a diet consisting solely of Artemia often developed this condition (Greco, 2004).

We now know that brine shrimp (Artemia sp.) contains inadequate levels of calcium and an imbalanced ratio of calcium to phosphorus, making it unsuitable as a staple diet for seahorses and pipefish even when enriched (Greco, 2004).

Seahorses afflicted with soft plate syndrome exhibit shortened life spans, decalcification of their exoskeleton, and poor survival rate amongst their fry (Greco, 2004). Pregnant males face the greatest risk of soft plate disease. Seahorse fry are known to incorporate calcium provided by their father into their own skeletons during their embryonic development, so when a pregnant male is deficient in calcium, his rapidly growing offspring typically suffer high mortalities due to the shortage of calcium - a condition akin to rickets in human children.

Fortunately, this debilitating condition is easily prevented by providing seahorses with adequate levels of bio available calcium either in their diet or in the aquarium water itself.

Mminerals can be obtained by fish directly from the water., (Greco, 2004).

I have never heard of a case of soft plate in a seahorse kept in a reef tank that received Kalkwasser (calcium hydroxide) via an automatic doser or regular supplementation of bio available calcium. Nor have I even seen this condition in seahorses that received a stable diet of enriched frozen Mysis relicta.

Paraphrased from:

Giwojna, Pete.
Unpublished.
Complete Guide To the Greater Seahorses in the Aquarium

Appendix

Setting up a hospital tank.

When a seahorse becomes ill it will often be necessary to medicate outside the main tank. Many medications are either toxic to the bio filter, the animals in the live rock, or the invertebrates, or even, with prolonged exposure, to the seahorses themselves.

For short-term dips and baths, a small container such as a new, rinsed ten litre bucket can be employed. However, this does not afford a great deal of visibility for observation, and for prolonged immersion treatments a hospital tank is essential. This does not need to be set up permanently. It can be dismantled when not in use and the filter removed and kept cycled by placing the inlet and outlet in a bucket of saltwater (or for sponge filters, the whole filter) and using a bottled ammonia product designed for feeding aquarium bio filters. Ammonia should never be used in the same water as livestock. If using a canister or other power filter in a smaller tank, adjust the flow rate appropriately so the seahorse is not having difficulty swimming or hitching. A cycled filter should be used unless the medication is bio filter toxic, as with certain antibiotics and medications such as Methylene blue.

The smaller the tank, the quicker the water quality will degrade, especially if there is no operational bio filter. Depending on the size of the seahorses and how many are being treated at once, ten gallons is a good benchmark. It is best not to use substrate, as it is easier to see and suction out any uneaten food, and monitor whether the fish is voiding feces, which indicates whether or not it is eating if this is not obvious.

Some diseases are untreatable presently, and if the fish is diagnosed as having contracted such a disease, the tank will need to be stripped and bleached and the seahorse euthanized. For that reason it is also a good idea not to use live rock.

Plastic hitches and heaters can be bleached, and airlines and filter media disposed of.

To prevent shock water in the hospital tank must be temperature matched to within one degree and PH matched to within 0.2 of the measurements from the tank the seahorse is being moved from. The new water used must be mixed with a powerhead and aged for twenty-four hours prior to use. Ideally this should also be aerated for twenty-four hours prior to adding the salt mix, but this stage can be skipped if the water is needed in an emergency.

A heater or chiller should be used to maintain the desired temperature if large swings in temperature would otherwise occur over short spaces of time.

The hospital tank should also have a thermometer, an airline and hitches. Test kits for ammonia, nitrite and PH (or better, a probe for reading PH) are essential. Detectable ammonia or nitrite should be addressed immediately with a large water change as even low level ammonia or nitrite toxicity will further complicate matters. In an emergency, Amquel+ can be used to neutralize ammonia and nitrite while new salt water is prepared.

Water Parameters

What you test for, why you test for it, and what the acceptable values are.

What	Optimum	When	Information
Ammonia (NH3NH4+)	0.00	Daily	Ammonia is the primary enemy of invertebrates & fish, capable of causing death in very low concentrations. Causes include: Immature filter, overfeeding, overstocking & dead stock.
Nitrite (N02)	0.00	Daily	Even trace levels of nitrite can destroy a well-presented invertebrate aquarium & cause fish much distress.

Nitrate (N03)	below 15 ppm	Weekly	A good overall indicator of general water quality & one that should be kept extremely low if invertebrates are to thrive. Constantly high nitrate levels usually reflect high fish stocking ratios.
Salinity	1.021 - 1.024	Daily	Salinity measures the total amount of dissolved solids in seawater. It is usually recorded as specific gravity (S.G.) But can also be referred to as parts per thousand (ppt). Constant evaporation of freshwater from the aquarium causes the salts to become more concentrated & the salinity to rise. To maintain stability regular addition of freshwater is needed.

pH	8.1 - 8.3	initially daily, then weekly	pH is a measure of the alkalinity or acidity of aquarium water. Some natural changes are to be expected during the day. Aquarium water could drop to as low as 7.9 at the end of the night, & peak at around 8.4 just before lights out. These natural pH cycles are gradual & tend not to stress livestock.
Phosphates (PO4)	0.00	Monthly	Invertebrates do not prosper when levels of phosphate get high. Phosphates arrive in the aquarium through unfiltered water, poor quality carbon & marine salts, but mostly through fish waste products. Nuisance algae will thrive where phosphate levels are high. High-quality water changes or phosphate removing resins can help alleviate the problem.

Dissolved Oxygen (02)	6 - 7 ppm	Monthly, or on demand	Both fish & invertebrates benefit from high levels of dissolved oxygen. Good water circulation is the key, as oxygen is drawn mainly from air & water. Dissolved oxygen also affects pH.
Copper	0	Intermittently, then on demand	Copper-based medications have proved very reliable in the treatment of various fish diseases. It is highly toxic to invertebrates & should never be used in aquarium housing these animals. Copper adversely affects seahorses. Copper can be introduced to the marine aquarium by way of domestic water & this should be tested from time to time.

Calcium	350 - 400 ppm	Monthly	Calcium is a vital element in the marine aquarium. A host of invertebrates draw it from the surrounding water and calcium reserves need to be replenished on a regular basis. Regular water changes usually achieve this. A reef tank may require the addition of biologically available calcium to keep levels optimum.
Carbonate Hardness (KH)	7dKH	Monthly	KH is a measurement of various carbonates & bicarbonates of calcium & magnesium, & borates in seawater. A stable KH will prevent rapid declines in alkalinity & subsequent drops in pH. Boosting the KH of aquarium seawater to between 12 - 18dKH using a proprietary generator has been recommended. However, left to their own devices, most aquariums settle

naturally to around 7dKH & there appears to be no advantage in constantly increasing dKH to unnatural levels.

Quarantine

There is absolutely no argument that prevention is better than cure. Yet one of the most common mistakes amongst experienced and inexperienced seahorse keepers alike is an abbreviated or non-existent quarantine protocol.

The most effective way to prevent seahorses from becoming ill is to ensure, as far as is humanly possible, that any new additions, particularly syngnathids that can be vectors for genus or species-specific pathogens, are disease free before being placed in the tank.

It is particularly important with wild caught seahorses, who may be carrying contagious diseases which take a while to become apparent and are currently as good as untreatable (e.g., Microsporidians such as *Glugea heraldi*, and certain bacterial diseases) and may result in having to strip down the display tank, sterilize the contents and in the worst case scenario, euthanize all the seahorses. Where people have been successful with wild caught without quarantine it is down to luck rather than judgement.

Even captive bred seahorses should be quarantined, though because of the much-diminished possibility of ectoparasites with long life cycles, the period can be shorter. Shipping is stressful, and captive bred seahorses can display symptoms of shock for a couple of days after arrival, such as lying on the substrate or reduced eye movement/response to stimulus, and often refuse frozen food for a short period of time after transportation. For this reason, it is important to have appropriately sized live foods on hand whether the seahorses are captive bred or wild caught.

There are several other important reasons to quarantine captive bred seahorses:

• Monitoring how much and what the new seahorses are eating and that they are not being out competed for food or stressed immediately after shipping by other fish.

• Ensuring they have not picked up disease from a holding tank if purchased from a local fish shop (if they have shared tanks with wild caught they should be quarantined as if wild caught themselves – with the exception of prophylactic treatment with anthelmintics - to remove any doubt – sometimes wild caught seahorses are even passed off as captive bred by unscrupulous dealers)

• Captive bred are sometimes shipped very young and may require smaller "grow out" tanks to ensure they thrive

Wild caught should be free of any occurrence of disease for six to eight weeks before being placed in the display tank - captive bred for three to four weeks (unless longer "grow out" is required as mentioned above).

Setting up a quarantine tank

1. Filtration

Buying new seahorses requires a degree of planning in advance. An active, cycled bio filter is required – pristine water quality is essential as detectable ammonia and nitrite is immunosuppressive even at the lowest detectable levels – and this takes time to cycle. As long as the filter will handle the bio load, the exact nature of the filtration is not overly critical, and the nitrates that build up due to the bio filtration process can be kept down with water changes.

Please refer to the website for instructions if necessary. A canister filter with an appropriate flow rate (i.e. around 3-6x tank turnover per hour) can be cycled before the tank is set up, by placing the inlet and outlet into a clean bucket that has not been used for cleaning products such as detergents, and using a bottled ammonia product aimed at feeding filtration bacteria in aquaria.

Mature live rock can be used at 1-2 pounds per gallon to aid bio filtration but this is only recommended with captive bred seahorses because of the previously mentioned issue with potential for untreatable diseases appearing with wild caught, requiring sterilization (although sterilized "dead" rock can be re-seeded and then used again unless medications such as copper have been used, which are absorbed into the rock and then released into the water). Seeded "dead" rock can of course be utilized if required.

Properly rated ultraviolet sterilizers can help control free swimming protozoan parasites and bacterial pathogens, but should be turned off if it becomes necessary to treat with water borne medications that may be inactivated by the ultraviolet light. This also applies to skimmers and carbon filter media, which can remove medications from the water but are otherwise safe to employ on quarantine tanks.

Inlets for equipment should be protected to prevent the seahorses being sucked into the filtration equipment or stuck to the tubing (especially with smaller species). This can be achieved by using tank dividers if necessary. If heaters are used they should be covered with heater guards or placed behind dividers so that the seahorses do not hitch to them, as this sometimes leads to burns, particularly to the pouch of male seahorses. Burns can be easily confused with bacterial infections, so care must be taken. A heater guard will remove any ambiguity as to the cause of the symptom.

2. Substrate

It is preferable not to use substrate for several reasons – it is easier to monitor whether the seahorses are defecating, and therefore eating, it is easier to dismantle and/or sterilize the tank after the quarantine, and live foods such as amphipods can burrow into the substrate and escape from the fish.

3. Tank size

The quarantine tank should be of an appropriate size for the seahorses, based on age and size, species and number of fish to be quarantined. It is not necessary to have a tank that is three times the height of the seahorse from coronet to uncurled tail tip for quarantine – this is a commonly quoted rule of thumb for breeding purposes because of the vertical copulatory rise, which is not a concern, and tall narrow tanks have less potential for gas exchange as there is less surface area.

Tanks that are too small for the species in question should be avoided, not only because of potential water quality issues but also because of the spatial requirements of the seahorses.

Hitches should be provided and an open airline for aeration if necessary.

Environmental parameters (ammonia, nitrite, PH, SG, temperature, and nitrate) should be monitored frequently.

General lab procedures

Accurate diagnosis is essential to the treatment of seahorse diseases. Access to a laboratory or a microscope can be greatly beneficial. When using a laboratory it should be made clear that the specimens originate from a fish. This is not only to advise them as to the methods they should use, but also to prevent an accidental public health alert.

Laboratories specializing in veterinary services are preferred over hospital or general commercial labs. Non-veterinary labs can perform adequately if they are knowledgeable and willing to make the necessary accommodations. Bacterial cultures should be done at 25 C or the appropriate temperature for the pathogen. Attention should be paid to communicating with the lab so the appropriate culture media for the suspected organism can be used. Depending on the type of media used, salt supplementation may be necessary.

In choosing a microscope it is important to have at least a 10X (low power), 40X (high dry) and 100X (high power, oil immersion) objectives. Through the use of a standard 10X eyepiece magnifications of 100X, 400X and 1000X can be achieved. While many infectious agents can be identified at 100 or 400 X, it takes 1000X to read bacterial gram stains and speciate many protozoans. When buying a microscope, price generally equates with quality.

In addition to a microscope some general laboratory supplies may be needed.

<u>Supplies</u>

Slides and cover slips
Scalpel and blades
Denatured ethyl alcohol
Sterile gloves
Culture swabs
Gram stain kit
eye dropper or pipette
Safety glasses
Scissors and forceps
cutting board
formalin
small containers
Camera with macro lens
Ruler

Slides and cover slips are necessary to prepare wet preps of currettings and biopsy material. They are also used in the preparation of a gram stain. Plastic cover slips are easier to work with but offer less visual clarity than glass.

A scalpel with disposable sterile blades is useful for autopsies, biopsies and currettings. Denatured ethyl alcohol is used for sterilizing the outer surface of autopsy specimens in preparation for internal culture. It can also be used to sterilize instruments by soaking. Care must be taken to reduce fumes by using a tightly sealed container at all times for such purposes. In addition, the risk of fire or environmental exposure must also be considered.

Sterile gloves come in handy if one wishes to do intraperitoneal cultures on autopsy specimens. The sterile paper packaging that the gloves are wrapped in is also useful to create a sterile field for such procedures. If one is not going to perform cultures a non-sterile glove is recommended for protection when handling specimens.

Sterile culture swabs with transport media are essential for getting viable specimens to the lab. Specialized varieties exist for viruses, aerobic bacteria, anaerobic bacteria and fungi. The laboratory usually supplies these free of charge but must be notified what type of cultures are to be performed. For seahorses, aerobic swabs seem to be the most useful. Cultures are transported as soon as possible in appropriate media to the lab at room temperature.

Mycobacterial cultures are best taken by small tissue samples sent on sterile saline moistened gauze as soon as possible without preservative. They can take a long time and/or are very expensive. As a cost/time saving measure, it may be preferred to identify the bacteria on tissue sections with the appropriate acid fast stain and histologic picture. The same holds true for Nocardia. However, Nocardia can be identified through Gram stain, by its characteristic right angle branching. Because of the difficulties involved, identification of these organisms may be out of the reach of many hobbyists.

A Gram stain is a quick easy procedure to do. Since splashing can occur it is important to wear eye protection. The dyes can stain clothing and objects, so an appropriate area must be chosen for this activity. It is also important to follow directions carefully.

A few pairs of stainless steel surgical scissors of different sizes are a good investment, especially if one plans to do post mortem examinations. A pair of toothed forceps is useful to manipulate specimens and hold small biopsies.

Necropsy specimens require an easily cleaned area for examination. A dedicated plastic kitchen cutting board can be especially useful in this function.

Submission of a biopsy specimen or autopsy tissues for histologic examination, requires quick tissue fixation. Freezing is not recommended. Refrigeration will help but it is better to place biopsy tissue in 10% formalin. Dilution of 33% hobbyist formalin with the appropriate amount of fresh water can be used to create a 10% solution. Enough fixative should be prepared to cover the specimen. An appropriately sized, leak proof container for the specimen must be found for transport to the laboratory. Small glass and plastic screw top containers work well for small specimens. Ziploc bags can also be used for both small and large specimens. Culture material should be taken before fixation since formalin will kill bacteria present in the specimen. In lieu of formalin any type of alcohol can also be used.

Fresh specimens sent for autopsy with culture should be refrigerated in a sealed plastic bag without water until shipped. For shipping, a Styrofoam box with a chill pack sent Next Day air works best. Documentation of lesions and abnormalities is best done with a digital camera. Macro function is essential since many of the abnormalities are very small. Good photographs can be used to analyse findings at a later date or to communicate with other hobbyists and experts as well as taking photos of the family picnic. When taking photographs it is helpful to place a ruler in the picture and labels as appropriate.

Good lighting and graphics software should also be considered in order to obtain quality photographs.

Through biopsy, culture and necropsy of tissues or autopsy material many common diseases of seahorses can be accurately diagnosed. Interested hobbyists need to decide which procedures they have the appropriate supplies and equipment to perform. Samples taken from living animals should be done only under anaesthesia. Gill biopsies are not recommended on living seahorses and needle biopsies of internal organs are best left to the experts. Evaluation of external lesions, swabs and scrapings can be performed. However the hobbyist must be knowledgeable of these procedures and be willing to deal with potential complications such as secondary bacterial infections. Consideration should also be paid to the condition and suffering of the patient. While one should never lose hope, sometimes euthanasia is the best option for those suffering who are beyond recovery.

Wet Prep

A wet prep is very useful for identifying protozoans, many parasites and some bacteria. Wet preps can be made from autopsy material or living specimens. Since many of the pathogens, especially protozoans do not survive for long off the host or without a living host, wet preps from living specimens are preferred. To prepare a wet prep one needs the following

slides
cover slips
freshly made salt water
scalpel
sterile swab
eyedropper or pipette
microscope

Samples can be obtained with a sterile swab or by scraping the affected area with a scalpel. Swabs work better for sampling larger areas or when scraping would cause too much damage. After using the appropriate implement the sample is gently smeared across the slide. A drop of salt water is added and a cover slip placed. The specimen can be immediately viewed under the microscope for analysis. Protozoa, parasites and their eggs, motility in Vibrio bacteria, are a few things that can be observed using this method.

Fixed slide

Material obtained with a swab or by scraping can be smeared on a slide, sprayed with fixative, allowed to air dry, then be evaluated under a microscope. Aerosols high in alcohol such as "White Rain" hairspray can be used as a spray fixative. The dried, fixed slide can be viewed unstained or could be stained at a later time. The downside to the fixed prep is that while it preserves the integrity of the specimen, organisms if present, are killed and motility cannot be observed.

Autopsy

Performing autopsies can not only aid in finding out more about the diseases of seahorses but can give one a better understanding of their anatomy and physiology. In addition, useful information such as height, weight and the general appearance of the internal organs can be obtained without the addition cost of laboratory evaluation.

First, an appropriate area must be found and the decision made as to what types of specimens if any, are to be obtained. Surface cultures and swabs for wet prep should be done on the specimen as soon as possible. External pictures, weights and measures should also be done at this time. Sampling of external lesions for histologic analysis can be done at this time unless peritoneal cultures are planned, which should be performed first.

Peritoneal samples for culture should be done in a sterile field, with sterile instruments and gloves. In order to get the best possible uncontaminated specimen, soak the deceased in alcohol for 5 minutes before the procedure. An incision is made along the keel just big enough to insert a culture swab but not too deep as to damage or puncture internal organs. Great care is taken to avoid the sides of the incision when taking the culture.

After the culture is taken and appropriately labelled, the incision is extended upward using scissors, along the neck, to the anterior part of the operculum. From the same starting point, a similar incision is extended downwards, posterior and up among the spine to meet with the previous incision at the operculum. The entire body wall with opercula can now be carefully lifted off to reveal the internal organs from the gills to the anal opening.

Some attachments to the peritoneum may remain but can be easily clipped with scissors. For males, the pouch can be opened in a similar fashion.

At this time it is useful to take photographs and make notes of the internal anatomy for later reference. The organs can now be removed, examined more closely and submitted for histologic analysis or wet prep.

Plates

Due to the format of the printing process, colour plates are not possible in this book. However, it would be a great pity to miss out on providing some of the spectacular colour photographs obtained through the autopsy process.

What we have done is to work around the problem, and provide the colour 'plate' on the rear cover of the book.

Figure 1:   The internal organs of a male seahorse. The labelled seahorse can be found in the black and white pages under the anatomy section on page 25 in this book.

Figure 2:   The internal organs of a female seahorse. The labelled seahorse can be found in the black and white pages under the anatomy section on page 27 in this book

Figure 3:   This is a microscopic section of a healthy kidney. It is important to see what a healthy kidney looks like to understand what is presented in a diseased kidney.

Figure 4:   This is a microscopic section of a kidney, which is infected with vibrio.

Figure 5:   This is a microscopic section of a liver, which has large fatty deposits, as in the health issue 'fatty liver disease'

Bibliography

Austin, D.; Austin, B., Bacterial Fish Pathogens: Diseases of Farmed and Wild Fish, (1999), New York, Springer-Verlag

Bassleer, G., Diseases in Marine Aquarium Fish, (2000), Hollywood Import & Export, Inc.

Henry, J., Clinical Diagnosis and Management by Laboratory Methods, (1996), Philadelphia, PA,: W.B. Saunders

Lom, J., Protozoan Parasites of Fishes, (1992), Elsevier Science

Noga, E., Fish Disease: Diagnosis and Treatment, (2000), Ames, Iowa, : Iowa State Press

Reichenbach-Klinke H. H., Fish Pathology, (1973), Neptune City, NJ. : T.F.H. Publications, Inc.

Roberts, R., Fish Pathology, (2001), Philadelphia, PA, : W.B. Saunders

Stoskopf, M., Fish Medicine, (1993), Philadelphia, PA, : W.B. Saunders

Untergasser, G., Handbook of Fish Diseases, (1990): Neptune City, NJ : TFH Publications, Inc.

Volk, W., Essentials of Medical Microbiology, (1991), Philadelphia, PA, : Lippincott

Lim, C, Webster, C., Nutrition and Fish Health, (2001), Binghamton, NY: Harworth Press

Herwig, N., Handbook of Drugs and Chemicals Used In The Treatment of Fish Diseases, (1979), Springfield, Ill. Charles Thomas Publications.

INDEX

W

Worms